INTERNATIONAL CENTRE FOR MECHANICAL SCIENCES

COURSES AND LECTURES - No. 120

RASTKO STOJANOVIĆ †
UNIVERSITY OF BELGRADE

NONLINEAR THERMOELASTICITY

LECTURES HELD AT THE DEPARTMENT
OF MECHANICS OF SOLIDS
JULY 1972

SPRINGER-VERLAG WIEN GMBH 1972

Originally published by Springer-Verlag Wien-New York 1972

ISBN 978-3-211-81200-6 ISBN 978-3-7091-2856-5 (eBook)
DOI 10.1007/978-3-7091-2856-5

Preface

I am mostly indebted to the authorities of CISM, and in particular to the Rectors and to the Secretary General for giving me the opportunity to give this course of six lectures on non-linear thermoelasticity. Having some points of view in this field of research which are not common in the general theories of thermo-mechanics, this is for me a good opportunity to develop systematically the argument and to compare different approaches to the problem of thermal stresses. The available time and space do not permit me to enter into many interesting problems of non-linear thermomechanics and I restricted the considerations only to the derivation of the constitutive relations and to their approximation by relations involving the second-order terms.

My colaborators and former students, Miss S. Milanovic and Dr. D. Blagojevic helped me in the preparation of this manuscript and I appreciate their assistence very much.

April 24, 1972 *R. Stojanovic*

Belgrade

Introduction

In the contemporary mechanics it is impossible to consider mechanical phenomena indipendently of the laws of thermodynamics even when the idealized processes like elasticity are considered. In thermoelasticity this coupling of mechanical and thermal fields becomes only accentuated. In the classical, linear thermoelasticity the influence of temperature is introduced into the stress-strain relations through the thermal strain tensor and the so obtained Neumann-Duhamel law simply replaces Hooke's law. If a linear heat conduction law, say Fourier's law, is added to the stress relations, the system of constitutive equations is completed. For the majority of engineering applications the linear theory gives sufficiently good results.

The development of the non-linear continuum mechanics automatically rises the problem of a general approach to thermo-mechanical phenomena. The fundamental set of equations is again represented by the heat conduction law and the stress relation, and the main restrictions upon these relations are imposed by the laws of thermodynamics. However, the energy functions which appear in such general formulations are considerably more complicated then in the linear theory.

Without entering into details of the development

of the non-linear thermoelasticity we shall mention here some of the most important contributions. In the already classical work on finite thermoelastic deformations A. SIGNORINI [35] gives a fine presentation of the first achievements in this direction. One of the first problems investigated by the pioneers was the dependence of the elastic moduli on temperature (BRILLOUIN [3] , [4] , [5] , [6] , [7] , SIGNORINI [34] , [35]) and the problem of energy (TOLOTTI [43]). A survey of the principal results up till 1952 is given by TRUESDELL [47] in his excellent critical review of the nonlinear elasticity and fluid dynamics.

Most of the modern approaches to the non-linear thermoelasticity are related closely to the development of the non-linear elasticity. In this respect we mention here the energetic considerations of GREEN and ADKINS [19] which are significant for the consideration of temperature and not entropy as an independent variable.

Independent attempts, based on the generalization of the isothermal non-linear elasticity, were made by JINDRA [26] and DILLON [17] . Dillon particularly investigated a second-order approximation. Second-order theories were also treated by HERRMANN [25] ; CHAUDHRY [10] and JOHNSON [25] A brief review of some trends in the non-linear thermoelasticity was given in 1964 by NOWACKI [28] .

A general theory of thermo-mechanics was developed during the last fifteen years and this development is closely connected with the names of GREEN, RIVLIN, TRUESDELL, COLEMAN, NOLL and others (see for instance references [8], [12] [13], [14], [16], [21], [48], [49], [50], [51] etc. For a survey of some of the latest developments see TRUESDELL [48])

However, most of the work done in thermo-mechanics was connected with viscoelasticity, not with thermoelasticity. This is quite natural since viscoelasticity is an irreversible process, and elasticity is a reversible and thus thermodynamically a much simpler process.

The choice of independent variables is not a simple problem (see e.g. GREEN and LAWS [22]). In the most of the above mentioned papers as independent variables appear entropy or internal energy. Since temperature is an observable and measurable quantity, we shall prefere temperature as the independent variable instead of the non-measurable quantities.

The approach to thermoelasticity which follows differs from the approaches of the mentioned authors, but it is not in contradiction with them. The basic observations upon which rests the theory to be developed here are:

1) Thermoelastic stresses, owing to non-uniform dilatations in a non-uniform temperature field, are produced by thermal strains. These strains don't satisfy the compatibility conditions. In a body at a uniform temperature there are no thermoelastic stresses. The classical thermoelasticity takes

care of the incompatibility of thermal strains, but some of the modern treatments disregard this fact.

2) The laws and principles of mechanics and thermodynamics should not be violated by any theory dealing with mechanical and thermal phenomena.

3) The coefficients of elasticity change with variations of temperature. While this fact was extensively treated by first investigators in nonlinear thermoelasticity, in some of the recent contributions this fact was also disregarded.

However, if the general theories of thermomechanical processes we regard deformation, temperature, temperature gradients etc. as independent variables without specifying how these quantities physically produce certain effects, a theory which is based more on the physical picture of a process should not be necessarily in contradiction to the general theories. This is actually the position of the theory to be developed on the following pages.

We assume that thermoelastic stresses are produced by certain incompatible strains and derive the constitutive relations from the laws of thermodynamics. The mathematical form of the energy equation has slightly to be modified in order to include incompatible deformations. The influence of temperature is assumed to be dual: through the thermal strains and through the variations of the coefficients of elasticity

Chapter 1

Thermo-Mechanical Preliminaries

Let $\mathcal{B}$ be a continuous material body at an instant t_0 of time, occupying in the space a configuration K_0 which will be considered as a reference configuration. We assume that in the reference configuration the body is free from stresses and that all points of the body are at the same reference temperature ϑ_0. In the configuration K_0 the points X of the body are referred to a system of coordinates X^K with the fundamental metric tensor G_{KL}, $K, L = 1, 2, 3$. The coordinates X^K will be called material coordinates, since the material points X of the body at any of its subsequent configurations may be identified through their coordinates X^K at the initial configuration.

At an instant t of time the body will be in its instantaneous configuration K_t and the points X of the body occupy the instantaneous positions x^k in the space. The equations

$$x^k = x^k(X^1, X^2, X^3; t) ,$$

(1.1)

$$X^K = X^K(x^1, x^2, x^3; t) ,$$

(considered as material constants).

Approximate constitutive relations are derived from the general relations for isotropic materials. The approximation is 'exact" within the frames of a second-order theory. The last section is dedicated to a brief survey of some results in the non-linear theory of polar elastic materials.

Owing to the restricted time, many interesting and important features of the non-linear theory will not be even mentioned. This is particularly connected with some second-order effects in the propagation of thermoelastic waves, for which we make only a reference to the work of JOHNSON [27] and CHADWICK and SEET [9] .

represent the equations of motion of the body. The coordinate system x^k may be chosen indipendently of the system X^K and we call x^k spatial coordinates (TRUESDELL [47]). The fundamental tensor corresponding to the system of coordinates x^k will be denoted by g_{kl} .

Let $\underset{\sim}{f}$, with components f^i , be the vector of force per unit mass acting upon the points of the body, and let $\underset{\sim}{t}$ be the Cauchy stress with components t^{ij} . For purely mechanical processes there are two fundamental laws of mechanics, the balance of the linear momentum and the balance of the moment of momentum.

Denoting by $\underset{\sim}{r}$ the position vector of the points of the body considered, for an arbitrary portion ϑ of $\mathcal{B}$, bounded by a surface a , the law of balance of the linear momentum reads

$$\frac{d}{dt}\int_{\vartheta}\rho\underset{\sim}{v}\,d\vartheta = \int_{\vartheta}\rho\underset{\sim}{f}\,d\vartheta + \oint_{a}\underset{\sim}{t}\,da\ , \tag{1.2}$$

where ρ is the density of mass at the points of the body in the instantaneous configuration K_t , and $\underset{\sim}{v} = \dot{\underset{\sim}{r}}$ is the velocity vector. The law of balance of the moment of momentum reads

$$\frac{d}{dt}\int_{\vartheta}\rho\underset{\sim}{r}\times\underset{\sim}{v}\,d\vartheta = \int_{\vartheta}\rho\underset{\sim}{r}\times\underset{\sim}{f}\,d\vartheta + \oint_{a}\underset{\sim}{r}\times\underset{\sim}{t}\,da\ . \tag{1.3}$$

In the componental form these two laws reduce to the fundamental laws of motion of Cauchy (*)

(1.4) $$\rho \dot{v}^i = \rho f^i + t^{ij}_{,j} ,$$

(1.5) $$t^{[ij]} \equiv \frac{1}{2} (t^{ij} - t^{ji}) = 0 .$$

The stress tensor cannot be determined from the equations of motion and it is necessary to introduce additional assumptions about behaviour of the material of the body considered. These assumptions are usually called principles.

(*) Throughout these lectures we use the notation of the theory of double tensor fields (cf. ERICKSEN [19] , with capital roman characters related to material coordinates and lower case roman characters related to spatial coordinates. If $T^{\cdots}_{\cdots}(\underset{\sim}{X}, \underset{\sim}{x})$ is a tensor field, $T^{\cdots}_{\cdots,m}$ denotes the partial covariant derivative when $\underset{\sim}{X}$ is kept constant, and $T^{\cdots}_{\cdots,M}$ is the partial covariant derivative when $\underset{\sim}{x}$ is kept constant. The total covariant derivatives are denoted by $T^{\cdots}_{\cdots;m}$ and $T^{\cdots}_{\cdots;M}$ with

$$T^{\cdots}_{\cdots;M} = T^{\cdots}_{\cdots,M} + T^{\cdots}_{\cdots,m} x^m_{;M} , \qquad T^{\cdots}_{\cdots;m} = T^{\cdots}_{\cdots,m} + T^{\cdots}_{\cdots,M} X^M_{;m}$$

where

$$x^m_{;M} = \frac{\partial x^m}{\partial X^M} , \qquad X^M_{;m} = \frac{\partial X^M}{\partial x^m} ,$$

are the deformation gradients.

For elastic bodies the principles may be formulated as follows:

Principles of determinism for the stress: The stress in an elastic body at an instant of time is determined by the instantaneous configuration K_t of the body.

Principle of local action: The stress at a given point X is determined by the configuration of an arbitrary small neighborhood of X, and the influence of the points outside this neighborhood may be disregarded.

According to these two principles, the constitutive equation for an elastic material will be

$$\underset{\sim}{t} = \underset{\sim}{t}(\underset{\sim}{F}) , \tag{1.6}$$

where $\underset{\sim}{F}$ is the matrix of the deformation gradients at the time t,

$$F^{k}_{\cdot K} = x^{k}_{;K} . \tag{1.7}$$

Principle of material frame-indifference (principle of objectivity): Constitutive equations must be invariant under changes of frame of reference.

If two motions, $\underset{\sim}{x} = \underset{\sim}{x}(X, t)$ and $\overset{*}{\underset{\sim}{x}} = \overset{*}{\underset{\sim}{x}}(X, \overset{*}{t})$ differ for an arbitrary rigid body motion,

$$\begin{aligned} \overset{*}{\underset{\sim}{x}} &= \underset{\sim}{c} + \underset{\sim}{Q} \cdot \underset{\sim}{x}(X,t) , \\ \overset{*}{t} &= t - a , \end{aligned} \tag{1.8}$$

where $\underset{\sim}{c} = \underset{\sim}{c}(t)$ is an arbitrary time-dependent vector, $\underset{\sim}{Q} = \underset{\sim}{Q}(t)$ a time-dependent orthogonal tensor, and a an arbitrary number, the two motions are equivalent. If we consider a process with the motion

$$\underset{\sim}{x} = \underset{\sim}{x}(X,t) \quad ,$$

and with constitutive equations

$$\varepsilon = \varepsilon(\underset{\sim}{F}, X, t) \; ,$$

$$\underset{\sim}{V} = \underset{\sim}{V}(\underset{\sim}{F}, X, t) \; ,$$

$$\underset{\sim}{T} = \underset{\sim}{T}(\underset{\sim}{F}, X, t) \; ,$$

where ε is a scalar field, $\underset{\sim}{V}$ a vector field and $\underset{\sim}{T}$ a second-order tensor field, according to the principle of objectivity the constitutive equations must be satisfied also for the motion given by (1.8)

$$\overset{*}{\varepsilon} = \varepsilon(\underset{\sim}{F}^*, X, t^*) = \varepsilon(\underset{\sim}{F}, X, t) \quad ,$$

$$\overset{*}{V}{}^i = \overset{*}{V}{}^i(\underset{\sim}{F}^*, X, t^*) = V^m(\underset{\sim}{F}, X, t)\, Q^i_{\cdot m}(t) \quad , \tag{1.9}$$

$$\overset{*}{T}{}^{ij} = \overset{*}{T}{}^{ij}(\underset{\sim}{F}^*, X, t^*) = T^{mn}(\underset{\sim}{F}, \underset{\sim}{X}, t)\, Q^i_{\cdot m}(t)\, Q^j_{\cdot n}(t) \quad .$$

Principle of equipresence: If a quantity appears as an independent variable in one constitutive equation, it should appear in all constitutive equations unless its presence is in contradiction with some laws of physics or principle of objectivity.

However, mechanical processes cannot be considered isolated from the sourroundings and one of the most important effects of this interaction is the exchange of energy. If $\varepsilon(X,t)$ is a <u>specific internal energy</u> and

$$E(\vartheta) = \int_{\vartheta} \rho\,\varepsilon\, d\vartheta \tag{1.10}$$

is the total internal energy of a portion ϑ of a body, the exchange of energy with the surroundings is governed by the <u>energy balance law</u>, or the first <u>law of thermodynamics</u>,

$$\frac{d}{dt}\int_{\vartheta} \rho\left(\frac{1}{2}\underset{\sim}{v}\cdot\underset{\sim}{v} + \varepsilon\right)d\vartheta = \int_{\vartheta} \rho(\underset{\sim}{f}\cdot\underset{\sim}{v} + q)\,d\vartheta + \oint_{a}(\underset{\sim}{t}\cdot\underset{\sim}{v} + h)\,da\ , \tag{1.11}$$

where $q = q(X,t)$ is the specific heat production per unit mass, and $h = h^i n_i$ is the heat flux. In view of the equations of motion (1.4) and (1.5), the energy balance law (1.11) reduces to the local energy balance equation

$$\rho\dot{\varepsilon} = t^{ij}d_{ij} + h^i_{,i} + \rho q\ , \tag{1.12}$$

where d_{ij} is the rate of deformation tensor,

$$d_{ij} = v_{(i,j)} = \frac{1}{2}(v_{i,j} + v_{j,i})\ . \tag{1.13}$$

The energy balance law (1.11) establishes an interaction between the kinetic energy

$$T = \frac{1}{2}\int_{\vartheta} \rho\, \underset{\sim}{v} \cdot \underset{\sim}{v}\, d\vartheta$$

internal energy $E(\vartheta)$, rate of mechanical working

$$P = \int_{\vartheta} \rho \underset{\sim}{f} \cdot \underset{\sim}{v}\, d\vartheta + \oint_{a} \underset{\sim}{t} \cdot \underset{\sim}{v}\, da \ ,$$

and the non-mechanical energy production

$$Q = \int_{\vartheta} \rho q\, d\vartheta + \oint_{a} h\, da \ .$$

Only a part of the non-mechanical energy production is mechanically recoverable. If $\theta = \theta(X, t)$ is the temperature field and if $\eta = \eta(X, t)$ is the specific entropy, such that

$$H(\vartheta) = \int_{\vartheta} \rho \eta\, d\vartheta \ ,$$

where $H(\vartheta)$ is the total entropy of the portion ϑ of the body considered, the quantity

(1.14) $$\Gamma \equiv \dot{H} - \oint_{a} \frac{h}{\theta}\, da - \int_{\vartheta} \rho \frac{q}{\theta}\, d\vartheta$$

represents the production of entropy. The entropy inequality, called also the second law of thermodynamics, or the Clausius-Duhem inequality states that the entropy production is non-negative,

(1.15) $$\Gamma \geq 0 \ .$$

If γ is the specific entropy production,

$$\Gamma = \int_{\vartheta} \rho \gamma \, d\vartheta \tag{1.16}$$

the entropy inequality may be reduced to the local form

$$\gamma = \dot{\eta} - \frac{1}{\rho\vartheta} h^{i}_{,i} - \frac{q}{\vartheta} + \frac{1}{\rho\vartheta^{2}} h^{i} \vartheta_{,i} \geq 0 \ . \tag{1.17}$$

Summarizing the results we see that in the study of thermo-mechanical processes we have to consider simultaneously the following fields: deformation gradients $x^{k}_{;K}$, temperature ϑ , internal energy ε , specific entropy η , stress t^{ij} and heat flux h^{i} . The fundamental laws are: balance of linear momentum (1.2), balance of moment of momentum (1.3), balance of energy (1.11) and the entropy inequality (1.15).

If $\underset{\sim}{F}$ and ϑ are selected to be independent variables t , h , ε , and η have to be determined by constitutive relations. Constitutive equations, on the other hand, cannot be arbitrary functions of the variables, but have to satisfy the entropy inequality (cf. COLEMAN and NOLL [12] , COLEMAN [11]).

For later applications it is convenient to introduce the <u>specific free energy</u> ψ ,

$$\psi = \varepsilon - \vartheta\eta \ , \tag{1.18}$$

and the entropy inequality becomes

$$(1.19) \qquad -\rho\dot{\psi} - \rho\eta\dot{\vartheta} + \rho\dot{\varepsilon} - h^i_{,i} - \rho q + \frac{1}{\vartheta} h^i \vartheta_{,i} \geq 0 \ .$$

Assuming that heat never flows against a temperature gradients it follows that

$$(1.20) \qquad h^i \vartheta_{,i} \geq 0$$

and h^i cannot be independent of the temperature gradient. The relation

$$(1.21) \qquad h^i = h^i(\underset{\sim}{F}\ , \eta\ , \mathrm{grad}\,\vartheta)$$

represents the heat conduction law. PIPKIN and RIVLIN [31] demonstrated that $h^i(\underset{\sim}{F}, \eta, 0) = 0$, proving thus the non-existence of the piezo-caloric effect. The most frequently met form of the constitutive equation for the heat flux is

$$(1.22) \qquad h^i = K^{ij}(\underset{\sim}{F}, \eta)\,\vartheta_{,j} \ ,$$

where K^{ij} is the heat conduction tensor. WANG [50] proved that $\underset{\sim}{K}$ is a symmetric tensor. In the linear approximation it is assumed that $\underset{\sim}{K}$ is a constant tensor, and for thermally isotropic bodies $K^{ij} = \varkappa\, g^{ij}$.

If we write

$$3) \qquad \rho\vartheta\dot{\eta} = h^i_{,i} + \rho q \ ,$$

and if we assume $\eta = \eta(X, \vartheta)$ and put $\vartheta \frac{\partial \eta}{\partial \vartheta} = C$, it follows that

$$C \frac{\partial \vartheta}{\partial t} = \frac{1}{\rho}(K^{ij}\vartheta_{,j})_{,i} + q \quad , \tag{2.24}$$

which represents the equation of heat conduction.

Since h^i is a function of $\text{grad}\,\vartheta$, according to the principle of equipresence we must assume also that the remaining constitutive equations are of the form

$$t^{ij} = t^{ij}(x^k_{;K}, \vartheta, \vartheta_{,m}) \; ,$$

$$\psi = \psi(x^k_{;K}, \vartheta, \vartheta_{,m}) \; , \tag{2.25}$$

$$\eta = \eta(x^k_{;K}, \vartheta, \vartheta_{,m}) \; .$$

From (1.13), (1.17) and (1.19) we have

$$-\rho\dot{\psi} - \rho\eta\dot{\vartheta} + t^{ij}d_{ij} + \frac{1}{\vartheta}h^i\vartheta_{,i} \geq 0 \; , \tag{2.26}$$

and since

$$t^{ij}d_{ij} = t^{\cdot j}_{k}X^K_{;j}\dot{x}^k_{;K} \; ,$$

using (1.25) we obtain

$$\left(-\rho\frac{\partial\psi}{\partial x^k_{;K}} + t^{\cdot j}_{k}K^K_{;j}\right)\dot{x}^k_{;K} - \rho\left(\frac{\partial\psi}{\partial\vartheta} + \eta\right)\dot{\vartheta} - \rho\frac{\partial\psi}{\partial\vartheta_{,m}}\dot{\vartheta}_{,m} + \frac{1}{\vartheta}h^i\vartheta_{,i} \geq 0 \, . \tag{2.27}$$

This inequality must be satisfied for arbitrary rates and it follows that

(1.28) $$\frac{\partial \psi}{\partial \vartheta_{,m}} = 0 \quad ,$$

(1.29) $$\eta = - \frac{\partial \psi}{\partial \vartheta} \quad ,$$

(1.30) $$t^{j}_{k} = \rho \frac{\partial \psi}{\partial x^{k}_{;K}} x^{j}_{;K} \quad .$$

Hence we see that ψ , η and t^{ij} are not functions of the temperature gradients (COLEMAN and MIZEL (*) [13]).

(*) COLEMAN and MIZEL assumed that $\underset{\sim}{t}$ determined by (1.30) is a thermostatic stress, corresponding to $\dot{\vartheta} = 0$, grad $\vartheta = 0$, and that the total stress consists of $\underset{\sim}{t}$ and of a dissipative stress $_{D}\underset{\sim}{t}$ which is a function of grad ϑ and of velocity gradients $\dot{\underset{\sim}{F}}$. They have shown that in this case h^{i} may be also a function of $\dot{\underset{\sim}{F}}$. COLEMAN and NOLL 12 introduce also the dissipative part of the stress tensor, writing $t^{ij} = {}_{E}t^{ij} + V^{ijkl}(\underset{\sim}{F},\eta)d_{kl}$, where d_{kl} is the rate of strain tensor. This assumption is based on the fact that purely elastic stress, without viscosity, is inadequate for the description of processes in which dissipation effects of heat conduction and viscosity are of the same order of magnitude (e.g. propagation of sound waves in gases).

Chapter 2

Thermal Strains

In the following we assume that the distribution of temperature in a body is a known function of position and time,

$$\vartheta = \vartheta(X, t) . \tag{2.1}$$

It is well known from elementary physics that heated bodies change their dimensions. If a free body of length L at an initial temperature ϑ_0 is heated to a temperature $\vartheta = \vartheta_0 + T$, $T = \text{const.}$, its length will whange to $l = L(1 + \alpha t)$ where α is the coefficient of thermal dilatation of the material.

If T is a function of position, at each point X of a body an elementary length ΔL will become after heating $\Delta l = \Delta L[1 + \alpha T(X)]$. Let z^α be Cartesian coordinates. If coordinates of two points of a homogeneous isotropic body differ in the initial reference configuration for dZ^A , after heating this difference will be

$$\delta z^\alpha = \delta^\alpha_A (1 + \alpha T) \, dZ^A , \tag{2.2}$$

and the fundamental metric form becomes

(2.3) $$dS^2 = \delta_{AB}\, dZ^A dZ^B = \frac{1}{(1+\alpha T)^2}\delta_{\alpha\beta}\, dz^\alpha dz^\beta,$$

or

(2.4) $$ds^2 = \delta_{\alpha\beta}\, dz^\alpha dz^\beta = (1+\alpha T)^2 \delta_{AB}\, dZ^A dZ^B .$$

At the so-called room temperatures, $\vartheta \approx 300°\,K$, the order of magnitude of the coefficient α is 10^{-4} and a linear approximation of (2.3) and (2.4) is justified for the majority of materials at temperatures sufficiently below the melting temperatures. Thus we may write

(2.5) $$\begin{aligned} dS^2 &= (1 - 2\alpha T)\, \delta_{\alpha\beta}\, dz^\alpha dz^\beta , \\ ds^2 &= (1 + 2\alpha T)\, \delta_{AB}\, dZ^A dZ^B . \end{aligned}$$

The quantity

(2.6) $$\overset{T}{e}_{\alpha\beta} = \alpha T \delta_{\alpha\beta}$$

represents the well known <u>thermal strain</u> in linear thermoelasticity.

In general the tensor $\overset{T}{\underset{\sim}{e}}$ does not satisfy the compatibility conditions,

(2.7) $$\zeta^{\nu\mu} = \epsilon^{\alpha\lambda\nu} \epsilon^{\beta\mu\gamma} \partial_\alpha \partial_\beta \overset{T}{e}_{\lambda\gamma}$$

where $\zeta^{\nu\mu}$ is the incompatibility tensor, $\epsilon^{\alpha\lambda\nu}$ is the Ricci alternating tensor and ∂_α stands for the partial derivatives $\partial_\alpha = \partial/\partial z^\alpha$. The conditions of compatibility will be satisfied only if ϑ = const., or if it is a linear function of coordinates. We say that thermal strains represent incompatible strains.

The incompatibility of thermal strains means that the relations (2.2) are not integrable. The mapping

$$z^\alpha = z^\alpha(Z^1, Z^2, Z^3) \; , \tag{2.8}$$

which brings the body from its initial state into a thermally strained state with the deformation tensor

$$\overset{T}{c}_{\alpha\beta} = (1 - 2\alpha T)\,\delta_{\alpha\beta} \tag{2.9}$$

does not exist. Using the language of geometry we may say that the body with the fundamental form (2.5) is no more in the Euclidean space.

However, bodies observed in the Euclidean space cannot leave this space and we know from elementary experience that in this space exist bodies with nonuniform temperature distributions. This geometrical constraint, that a body must remain in the Euclidean space, induces the appearance of an additional strain $\overset{\varepsilon}{e}(X)$, such that the total strain $\underset{\sim}{e} = \overset{T}{\underset{\sim}{e}} + \overset{\varepsilon}{\underset{\sim}{e}}$ is a compatible strain, and the total deformation

$$c_{\alpha\beta} = \overset{T}{c}_{\alpha\beta} + 2\,\overset{T}{e}_{\alpha\beta} \tag{2.10}$$

is a Euclidean metric tensor. Consequently, mappings of the form (2.8), but corresponding to the total deformation exist.

If a body consists of loose elements $N(X)$ thermally strained elements will deform independently of one another according to (2.6) and if the elements in the initial configurations constituted a continuous body, after heating the continuity will not be preserved. Each element must suffer an additional deformation in order to reastablish the continuity of the body after heating.

The basic assumption is that the thermal strains $\overset{T}{\underset{\sim}{e}}$ do not produce stresses, and that stresses are produced by the additional strains $\overset{E}{\underset{\sim}{e}}$. The additional strains will be called <u>elastic strains.</u>

In general the relations (2.2) - (2.6) are valid only for small increments $T = \vartheta - \vartheta_0$ of temperature. A more general model may be introduced, valid for arbitrary temperature fields and properties of solids (STOJANOVIC, DJURIC and VUJOSEVIC [41]).

An element dX^K of a body in the initial configuration K_0 ($\vartheta = \text{const.}$), after heating of the body to a temperature $\vartheta = \vartheta_0 + T(X)$ becomes

$$du^\lambda = \theta_K^{(\lambda)}(X, T)\, dX^K, \tag{2.11}$$

where $\theta_K^{(\lambda)}$ are <u>thermal distorsions</u>. In general

$$2S_{KL}^{(\lambda)} \equiv \theta_{K,L}^{(\lambda)} - \theta_{L,K}^{(\lambda)} \neq 0 \qquad (2.12)$$

the relations (2.11) are nonintegrable, and u^λ cannot be interpreted as coordinates of the Euclidean space. If $\theta_{(\lambda)}^{L}$ are reciprocal thermal distorsions,

$$\theta_K^{(\lambda)}\theta_{(\lambda)}^{L} = \delta_K^L \quad , \qquad \theta_K^{(\lambda)}\theta_{(\mu)}^{K} = \delta_\mu^\lambda \ , \qquad (2.13)$$

where δ_K^L, δ_μ^λ are Kronecker delta symbols, e.g. $\delta_K^L = \begin{cases} 1, & K = L \\ 0, & K \neq L \end{cases}$ from (2.12) follows

$$dX^K = \theta_{(\lambda)}^{K} du^\lambda \ .$$

The line element

$$dS^2 = G_{KL}\theta_{(\lambda)}^{K}\theta_{(\mu)}^{L} du^\lambda du^\mu \qquad (2.14)$$

is no more a Euclidean line element. The tensor

$$\overset{T}{C}_{\lambda\mu} = G_{KL}\theta_{(\lambda)}^{K}\theta_{(\mu)}^{L} \qquad (2.15)$$

represents a deformation in a non-Euclidean space. The thermal deformation (2.15) brings the body from an initial configuration K_0 into a non-Euclidean configuration K .

Since non-Euclidean configurations of bodies initially in the Euclidean space are impossible, simultaneously with thermal distorsions at the points of the body appear <u>elastic distorsions</u> $\Phi_K^{(\lambda)}(X, T)$ such that

$$\theta^{K}_{(\lambda)}\Phi^{(\lambda)}_{k} = X^{K}_{;k} \,. \tag{2.16}$$

If $\Phi^{k}_{(\lambda)}$ are reciprocal elastic distorsions,

$$\Phi^{(\lambda)}_{k}\Phi^{k}_{(\mu)} = \delta^{\lambda}_{\mu} \;, \quad \Phi^{(\lambda)}_{k}\Phi^{m}_{(\lambda)} = \delta^{m}_{k} \;, \tag{2.17}$$

we have also

$$\Phi^{k}_{(\lambda)}\theta^{(\lambda)}_{K} = x^{k}_{;K} \,. \tag{2.18}$$

The elastic distorsions bring the body from a non-Euclidean configuration K into an instantaneous Euclidean configuration K_t , and $X^{K}_{;k}$ and $x^{k}_{;K}$ represent gradients of the <u>total deformation</u> from K_0 to K_t .

Let us expand the distorsions $\theta^{(\lambda)}_{L}(X,T)$ into a power series,

$$\theta^{(\lambda)}_{L}(X,T) = \theta^{(\lambda)}_{L}(X,0) + \alpha^{\lambda}_{\cdot L}T + \underset{1}{\alpha}^{\lambda}_{\cdot L}T^{2} + \ldots \tag{2.19}$$

If $T = 0$, (2.11) reduces to the identity transformation,

$$\theta^{(\lambda)}_{L} = \delta^{\lambda}_{L} \,. \tag{2.20}$$

$\alpha^{\lambda}_{\cdot L}$, $\underset{1}{\alpha}^{\lambda}_{\cdot L}$, ... are coefficients of thermal expansion, and for isotropic materials the coefficients are isotropic tensors,

$$\alpha^{\lambda}_{\cdot L} = \alpha\,\delta^{\lambda}_{L} \;, \quad \underset{1}{\alpha}^{\lambda}_{\cdot L} = \underset{1}{\alpha}\,\delta^{\lambda}_{L} \;, \ldots \tag{2.21}$$

For $T=0$ according to (2.16) - (2.18) and (2.20) the elastic distorsions become deformation gradients, but since the body suffers no deformations we have also

$$\Phi^k_{(\lambda)}(X,0) = \delta^k_\lambda . \tag{2.22}$$

If the functions $\Phi^k_{(\lambda)}(X,T)$ are also expanded into power series, we have now

$$\Phi^k_{(\lambda)}(X,T) = \delta^k_\lambda + \varphi^{\cdot k}_{\cdot\lambda}(X)\,T + \ldots \tag{2.23}$$

The conditions (2.20) and (2.23) will be identically satisfied if we put

$$\begin{aligned} \theta^{(\lambda)}_L &= \delta^\lambda_L + T\alpha^{\lambda}_{\cdot L}(X,T) , \\ \Phi^k_{(\lambda)} &= \delta^k_\lambda + T\varphi^{k}_{\cdot\lambda}(X,T) . \end{aligned} \tag{2.24}$$

If $T = \text{const.}$ in the whole body, the body suffers a homogeneous deformation. From (2.11) and $(2.14)_1$ follows then, if X^k are Cartesian coordinates.

$$u^\lambda = (\delta^\lambda_K + T\alpha^{\lambda}_{\cdot K})X^K , \tag{2.25}$$

and $T\alpha^{\lambda}_{\cdot K} = \omega^\lambda_{;K}$, where ω^λ are components of a displacement vector.

We assume that in a problem of thermoelasticity thermal distorsions are known if we know the distribution of temperature throughout the body. The elastic distorsions have to be determined from the equilibrium and boundary conditions, or from the equations of motion in dynamical problems.

Chapter 3
The Non-Euclidean Configuration K

The configuration K is inadmissible in the Euclidean space and we may assume that it belongs to a non-Euclidean, linearly connected space L_3, the geometry of which is determined by the thermal distorsions $\theta_L^{(\lambda)}$. The space L_3 should reduce to the Euclidean space when the linear differential relations (2.11) are integrable.

If $S_{LK}^{\cdot\cdot M}$ is a third-order tensor defined by the relation

$$S_{LK}^{\cdot\cdot M} = \theta_{(\lambda)}^{M} \partial_{[L} \theta_{K]}^{(\lambda)} , \tag{3.1}$$

according to (2.12), vanishing of $\underset{\sim}{S}$ represents the integrability condition for (2.11). However, the tensor $\underset{\sim}{S}$ may be interpreted as the torsion tensor of a linearly connected space with the coefficients of connection (cf. SCHOUTEN [32])

$$\Gamma_{LK}^{M} = \theta_{(\lambda)}^{M} \partial_L \theta_K^{(\lambda)} . \tag{3.2}$$

From (2.16) and (2.18) follows that we have also nonintegrable relations

$$du^{\lambda} = \Phi_l^{(\lambda)} dx^l , \tag{3.3}$$

with

$$2S_{lk}^{(\lambda)} = \partial_l \Phi_k^{(\lambda)} - \partial_k \Phi_l^{(\lambda)} \neq 0 \ . \tag{3.4}$$

If we define the coefficients of connection with respect to the elastic distorsions $\Phi_k^{(\lambda)}$,

$$\Gamma_{lk}^{m} = \Phi_{(\lambda)}^{m} \partial_l \Phi_K^{(\lambda)} \tag{3.5}$$

It could be easily vei , ed that Γ_{KL}^{M} and Γ_{Kl}^{m} are coeffi cients of connection of the same space L_3, but expressed in terms of two different coordinate systems.

The vectors $\underset{\sim}{\theta}^{(\lambda)}$ and $\underset{\sim}{\Phi}^{(\lambda)}$ represent in L_3 fields of parallel vectors and the corresponding Riemann-Chris toffel tensors identically vanish,

$$\begin{aligned} R_{NML}^{\cdots K} &= 2\,(\partial_N \Gamma_{ML}^{K} + \Gamma_{NT}^{K}\Gamma_{ML}^{T})_{[NM]} \equiv 0\ , \\ R_{nml}^{\cdots k} &= 2\,(\partial_n \Gamma_{ml}^{k} + \Gamma_{nt}^{k}\Gamma_{ml}^{t})_{[nm]} \equiv 0\ . \end{aligned} \tag{3.6}$$

Consequently, the space L_3 admits a symmetric covariant constant second-order tensor a_{KL}, or a_{kl},

$$\begin{aligned} \overset{\Gamma}{\nabla}_K a_{LM} &= \partial_K a_{LM} - \Gamma_{KL}^{N} a_{NM} - \Gamma_{KM}^{N} a_{LN} = 0\ , \\ \overset{\Gamma}{\nabla}_k a_{lm} &= \partial_k a_{lm} - \Gamma_{kl}^{n} a_{nm} - \Gamma_{km}^{n} a_{ln} = 0\ . \end{aligned} \tag{3.7}$$

The tensor $\underset{\sim}{a}$ is the fundamental (metric) tensor of the space L_3 and a_{LM} and a_{lm} are its non-holonomic components in the Euclidean space. If $a_{\lambda\mu}$ are its components with respect to the non-Euclidean coordinates u^{λ}, we may write

$$a_{LM} = a_{\lambda\mu}\theta_L^{(\lambda)}\theta_M^{(\mu)} ,$$

$$a_{lm} = a_{\lambda\mu}\Phi_l^{(\lambda)}\Phi_m^{(\mu)} . \tag{3.8}$$

and in analogy with the definitions of the material and spatial deformation tensors we may define the <u>material and spatial tensors of thermal deformation</u>, respectively, by

$$\overset{T}{C}_{LM} = a_{LM} , \qquad \overset{T}{c}_{\lambda\mu} = G_{LM}\,\theta^L_{(\lambda)}\theta^M_{(\mu)} , \tag{3.9}$$

and the <u>spatial and material tensors of elastic deformation</u>, respectively by

$$\overset{\varepsilon}{c}_{lm} = a_{lm} , \qquad \overset{\varepsilon}{C}_{\lambda\mu} = g_{lm}\,\Phi^l_{(\lambda)}\Phi^m_{(\mu)} . \tag{3.10}$$

Obviously the use of the words "material" and "spatial" is here only an analogy with the usual terminology.

The tensor a_{LM} may be obtained from $(3.7)_1$ by integration, with the initial condition that a_{LM} reduces to G_{LM} when $S_{LN}^{\cdot\cdot M}$ vanishes, i.e. when the increment of temperature vanishes.

For thermally isotropic materials we have seen

that the thermal distorsions for many materials may be approximated with sufficient accuracy with (2.2), i.e. with

$$(3.11)\qquad \theta^{(\lambda)}_{L} = (1+\alpha T)\,\delta^{\lambda}_{L} \qquad , \qquad \alpha = \text{const.},$$

The coefficients of linear connection are then

$$(3.12)\qquad \Gamma^{M}_{KL} = \frac{\partial}{\partial Z^{K}} \log(1+\alpha T) = \delta^{M}_{L} ,$$

and we assume that Z^K are Cartesian coordinates. The differential equations for the determination of a_{KL} read now

$$(3.13)\qquad \partial_K a_{LM} - 2\, a_{LM}\, \partial_K \log(1+\alpha T) = 0 ,$$

and the integrals are

$$(3.14)\qquad a_{LM} = \delta_{LM}(1+\alpha T)^2 .$$

When $T = 0$ we have $a_{LM} = G_{LM} = \delta_{LM}$. According to (3.9), a_{LM} represents the material tensor of thermal deformation, and the material tensor of thermal strain will be

$$(3.15)\qquad \overset{T}{E}_{LM} = \frac{1}{2}(\overset{T}{C}_{LM} - \delta_{LM}) = \frac{1}{2}(2\alpha T + \alpha^2 T^2)\,\delta_{LM} .$$

For small temperature increments T this reduces to the well known formula of linear thermoelasticity (2.6),

$$(3.16)\qquad \overset{T}{E}_{LM} \approx \alpha T \,\delta_{LM} .$$

In analogy with the usual expressions for strain and deformation tensors and in analogy with (3.8), (3.9) and (3.10) we introduce the following notation in which the label " E " means "elastic", the label " T " means "thermal, and the quantities without a label correspond to the total deformation $K_0 \to K_t$

$$\overset{-1}{\underset{T}{C}}{}^{LM} = a^{LM} = a^{\lambda\mu}\theta^L_{(\lambda)}\theta^M_{(\mu)} \quad , \qquad (3.17)$$

$$\overset{-1}{\underset{\varepsilon}{C}}{}^{lm} = a^{lm} = a^{\lambda\mu}\Phi^l_{(\lambda)}\Phi^m_{(\mu)} \quad , \qquad (3.18)$$

$$\overset{T}{E}_{LM} = \frac{1}{2}(\overset{T}{C}_{LM} - G_{LM}) \quad , \qquad (3.19)$$

$$\overset{T}{e}_{\lambda\mu} = \frac{1}{2}(a_{\lambda\mu} - \overset{T}{c}_{\lambda\mu}) = \overset{T}{E}_{LM}\theta^L_{(\lambda)}\theta^M_{(\mu)} \, , \qquad (3.20)$$

$$\overset{\varepsilon}{E}_{\lambda\mu} = \frac{1}{2}(\overset{\varepsilon}{C}_{\lambda\mu} - a_{\lambda\mu}) \quad , \qquad (3.21)$$

$$\overset{\varepsilon}{e}_{lm} = \frac{1}{2}(g_{lm} - \overset{\varepsilon}{c}_{lm}) \quad , \qquad (3.22)$$

$$\overset{T}{e}_{lm} = \overset{T}{E}_{LM}X^L_{;l}X^M_{;m} = \overset{T}{e}_{\lambda\mu}\Phi^{(\lambda)}_l\Phi^{(\mu)}_m \, , \qquad (3.23)$$

(3.24) $$\overset{\varepsilon}{E}_{LM} = \overset{\varepsilon}{E}_{\lambda\mu}\theta^{(\lambda)}_{L}\theta^{(\mu)}_{M} = \overset{\varepsilon}{e}_{lm}x^{l}_{;L}x^{m}_{;M} ,$$

(3.25) $$E_{LM} = \overset{\varepsilon}{E}_{LM} + \overset{T}{E}_{LM} ,$$

(3.26) $$e_{lm} = \overset{\varepsilon}{e}_{lm} + \overset{T}{e}_{lm} .$$

If $d\sigma$ is the element of arc in L_3 and if ds is the element of arc in the Euclidean space, according to $(3.8)_2$ and $(3.10)_2$ we have

(3.27) $$d\sigma^2 = a_{\lambda\mu}du^{\lambda}du^{\mu} = \overset{\varepsilon}{c}_{lm}dx^{l}dx^{m} ,$$

(3.28) $$ds^2 = g_{ij}dx^{i}dx^{j} = \overset{\varepsilon}{C}_{\lambda\mu}du^{\lambda}du^{\mu} .$$

Let c_i be the principal values of the tensor $\overset{\varepsilon}{\underset{\sim}{c}}$,

$$\overset{\varepsilon}{c}_{lm} = c\, g_{lm}$$

and let C_i be the principal values of the tensor $\overset{\varepsilon}{C}_{\lambda\mu}$,

$$\overset{\varepsilon}{C}_{\lambda\mu} = C a_{\lambda\mu} .$$

Then

$$d\sigma_i^2 = c_i\, ds_i^2$$

and

$$ds_i^2 = C_i\, d\sigma_i^2 \ .$$

Hence

$$c_i = \overset{-1}{C}_i \ . \qquad (3.29)$$

The principal invariants of the tensors $\overset{\varepsilon}{C}_{\lambda\mu}$ and $\overset{\varepsilon}{c}_{lm}$ are, respectively,

$$\begin{aligned} I^{\varepsilon}_{\underset{\sim}{C}} &= \delta^{\alpha}_{\lambda} \overset{\varepsilon}{C}{}^{\lambda}_{\alpha} = C_1 + C_2 + C_3 \ , \\ II^{\varepsilon}_{\underset{\sim}{C}} &= \frac{1}{2!} \delta^{\alpha\beta}_{\lambda\mu} \overset{\varepsilon}{C}{}^{\lambda}_{\alpha} \overset{\varepsilon}{C}{}^{\mu}_{\beta} = C_2C_3 + C_3C_1 + C_1C_2 \ , \\ III^{\varepsilon}_{\underset{\sim}{C}} &= \frac{1}{3!} \delta^{\alpha\beta\gamma}_{\lambda\mu\nu} \overset{\varepsilon}{C}{}^{\lambda}_{\alpha} \overset{\varepsilon}{C}{}^{\mu}_{\beta} \overset{\varepsilon}{C}{}^{\nu}_{\gamma} = C_1C_2C_3 \ . \end{aligned} \qquad (3.30)$$

and

$$\begin{aligned} I^{\varepsilon}_{\underset{\sim}{c}} &= \delta^{i}_{l} \overset{\varepsilon}{c}{}^{l}_{i} = c_1 + c_2 + c_3 \ , \\ II^{\varepsilon}_{\underset{\sim}{c}} &= \frac{1}{2!} \delta^{ij}_{lm} \overset{\varepsilon}{c}{}^{l}_{i} \overset{\varepsilon}{c}{}^{m}_{j} = c_2c_3 + c_3c_1 + c_1c_2 \ , \end{aligned} \qquad (3.31)$$

$$\underset{\sim}{III}^{\varepsilon}_{c} = \frac{1}{3!}\,\delta^{ijk}_{lmn}\,\overset{\varepsilon}{c}{}^{m}_{j}\,\overset{\varepsilon}{c}{}^{n}_{k} = c_1 c_2 c_3 \quad .$$

From (3.29) follows that between the invariants of the tensors $\overset{\varepsilon}{C}_{\lambda\mu}$ and $\overset{\varepsilon}{c}_{ij}$ are valid the same relations as in the isothermal theory of elasticity,

$$I^{\varepsilon}_{\underset{\sim}{c}} = \frac{II^{\varepsilon}_{\underset{\sim}{c}}}{III^{\varepsilon}_{\underset{\sim}{C}}} \; , \quad II^{\varepsilon}_{\underset{\sim}{c}} = \frac{I^{\varepsilon}_{\underset{\sim}{c}}}{III^{\varepsilon}_{\underset{\sim}{C}}} \; , \quad III^{\varepsilon}_{\underset{\sim}{c}} = \frac{1}{III^{\varepsilon}_{\underset{\sim}{C}}} \quad . \tag{3.32}$$

Chapter 4
Elastic Strains

According to the principle of determinism for stress and to the principle of local action, at an instant t of time the stress in elastic body at a point $\underset{\sim}{X}$ is determined by the relation of the form

$$\underset{\sim}{t} = \underset{\sim}{t}\,[\underset{\sim}{F}(\underset{\sim}{X},\, t)] \;, \tag{4.1}$$

where $\underset{\sim}{F}$ is the matrix of the deformation gradients corresponding to a deformation

$$x^i = x^i(\underset{\sim}{X},\, t) = x^i(X^1, X^2, X^3;\, t) \;. \tag{4.2}$$

This assertion is valid, however, only if the stress is produced by the total deformation (4.2), such that the deformation gradients result from the comparison of the initial and instantaneous configurations, K_0 and K_t .

From the analysis of the preceding sections it follows that, in the absence of volume forces, which are excluded from our considerations, stress is produced only by the elastic deformation, not by the total deformation. Therefore the constitutive equation for stress should read, instead of (4.1),

(4.3) $$\underset{\sim}{t} = \underset{\sim}{t}[\Phi^k_{(\lambda)}(\underset{\sim}{X}, t)] .$$

(STOIANOVIC, DJURIC, VUJOSEVIC [41] , [42]).

If we regard deformation gradients $F^k_{.K}$ as independent variables, equations of motion (4.2) may be obtained by integration of the exact differentials $F^k_{.K}\, dX^K$, i.e.

(4.4) $$x^k(\underset{\sim}{X}, t) = \int F^k_{.K}(\underset{\sim}{X}, t)\, dX^K ,$$

and the velocity field of the points of the body is determined by

(4.5) $$v^k(\underset{\sim}{X}, t) = \int \dot{F}^k_{.K}(\underset{\sim}{X}, t)\, dX^K .$$

In the case of thermo-elastic deformations we have from (2.18)

(4.6) $$x^k = \int \Phi^k_{(\lambda)}\, \theta^{(\lambda)}_K\, dX^K ,$$

and the velocity of the points $\underset{\sim}{X}$ at an instant of time t in the transition of the body from the configuration K_0 to K_t via K will be

(4.7) $$v^k = \int (\dot{\Phi}^k_{(\lambda)} \Phi^{(\lambda)}_l x^l_{;K} + x^k_{;L}\, \theta^L_{(\lambda)} \dot{\theta}^{(\lambda)}_K)\, dX^K$$

Since $x^k_{;K}\, dX^K = dx^k$, the term

(4.8) $$\dot{\Phi}^k_{(\lambda)} \Phi^{(\lambda)}_l = \overset{\varepsilon}{v}{}^k_{.l}$$

may be regarded as a generalized gradient of a velocity $\overset{\varepsilon}{v}{}^k$ with which a point $\underset{\sim}{X}$ passes from the configuration K to the configuration K_t.

If P and Q are two fixed and close to one another points in the (imagined) non-Euclidean configuration K of the body,

$$u^\lambda_Q - u^\lambda_P = du^\lambda$$

and if x^k_P and x^k_Q are the positions of these two points in the configuration K_t,

$$x^k_Q - x^k_P = dx^k ,$$

then

$$dx^k = \Phi^k_{(\lambda)} du^\lambda . \tag{4.9}$$

The difference $v^i_Q - v^i_P = dv^i$ of the velocities of these two points in the transition from K to K_t will be, with du^λ kept fixed,

$$d\overset{\varepsilon}{v}{}^k = \dot{\Phi}^k_{(\lambda)} \Phi^{(\lambda)}_l dx^l = \overset{\varepsilon}{v}{}^k_{.l}\, dx^l . \tag{4.10}$$

(cf. STOJANOVIC, DJURIC, VUJOSEVIC [42]). In general $\overset{\varepsilon}{v}{}^k_{.l}$ are not gradients and the relations (4.10) are not integrable.

Similarly

$$\overset{T}{V}{}^L_{.K} = \Theta^L_{(\lambda)} \dot{\Theta}^{(\lambda)}_K \tag{4.11}$$

measures the difference of velocities of two close to one-another points in the transition from the initial configuration K_0 to K ,

(4.12) $$d\overset{T}{V}{}^{L} = \overset{T}{V}{}^{L}_{.K}\, dX^{K} .$$

The tensor

(4.13) $$\overset{T}{v}{}^{k}_{.K} = x^{k}_{;L}\overset{T}{V}{}^{L}_{.K}$$

is the spatial representation $\overset{T}{V}{}^{L}_{.K}$.

Owing to the non-integrability of (4.10) and (4.12) the velocities $\overset{\varepsilon}{v}{}^{k}$ and $\overset{T}{V}{}^{L}$, or $\overset{T}{v}{}^{k}$, where

(4.14) $$d\overset{T}{v}{}^{k} = \overset{T}{v}{}^{k}_{.K}\, dX^{K}$$

may be only locally determined and the fields $\overset{\varepsilon}{\underset{\sim}{v}}(\underset{\sim}{X}, t)$ and $\overset{T}{\underset{\sim}{V}}(\underset{\sim}{X}, t)$ do not exist. However, the field

(4.15) $$v^{k}_{;L} = \overset{\varepsilon}{v}{}^{k}_{.L} + \overset{T}{v}{}^{k}_{.L} = \dot{x}^{k}_{;L}$$

is integrable, as follows from (4.6) and (4.7), and v^{k} is the field of effective velocities of points of the body.

The locally defined quantities $\overset{\varepsilon}{\underset{\sim}{v}}$ and $\overset{T}{\underset{\sim}{V}}$ we shall call <u>elastic</u> and <u>thermal velocities</u> (BLAGOJEVIC [2]).

The laws of balance of momentum and moment of momentum are independent of the nature of stress, and we assume these two laws in the form (1.2) and (1.3), with the corresponding laws of motion (1.4) and (1.5). However, the law of balance of energy has to be slightly modified. The work done by the thermoelastic stress $\underset{\sim}{t}$ is not done on the total deformation, but only on the elastic deformation and we rewrite the energy balance law in the form

$$\frac{d}{dt}\int_{\vartheta}\rho\left(\frac{1}{2}v^i v_i+\varepsilon\right)d\vartheta=\oint_a\left(t^{ij}\overset{\varepsilon}{v}_i+h^j\right)da_j+\int_{\vartheta}\rho\left(f^i v_i+q\right)d\vartheta . \tag{4.16}$$

Since the field $\overset{\varepsilon}{v}_i$ doesn't exist, the divergence theorem cannot be applied to the integral

$$\oint_a t^{ij}\,\overset{\varepsilon}{v}_i\,da_j .$$

However, for the incompatible deformations there is a modified divergence theorem (cf. Appendix A5 to [40]): If t^{ij} is a regular tensor field in ϑ and on a and if $\overset{\varepsilon}{v}_i$ is an "incompatible" velocity, then

$$\oint_a t^{ij}\overset{\varepsilon}{v}_i da_j = \int_{\vartheta}\left(t^{ij}_{,j}\overset{\varepsilon}{v}_i + t^{j}_{i}\,\dot{\Phi}^i_{(\lambda)}\Phi^{(\lambda)}_j\right)d\vartheta . \tag{4.17}$$

Introducing (4.17) into (4.16) and using the equations of motion (1.4) we obtain the local form of the energy equation,

$$\rho\dot{\varepsilon} = t^{ij}g_{il}\dot{\Phi}^l_{(\lambda)}\Phi^{(\lambda)}_j + t^{ij}_{,j}\left(\overset{\varepsilon}{v}_i - v_i\right) + h^j_{,j} + \rho q . \tag{4.18}$$

According to (4.15) we may assume that

$$\underset{\sim}{v} - \overset{\varepsilon}{\underset{\sim}{v}} = \overset{T}{\underset{\sim}{v}} , \tag{4.19}$$

and $t^{ij}_{,j}\overset{T}{v}_i$ represents the part of mechanical working which goes into the non-mechanical energy. If we put

$$(4.20) \qquad - t^{ij}_{,j} \overset{T}{v}_i = \rho q^*$$

and

$$(4.21) \qquad q^* - q = \tilde{q} ,$$

the total non-mechanical rate of working per unit volume of the body is $h^j_{,j} + \rho \tilde{q}$. The energy equation now reads

$$(4.22) \qquad \rho \dot{\varepsilon} = t^{ij} g_{il} \dot{\Phi}^l_{(\lambda)} \Phi^{(\lambda)}_j + h^j_{,j} + \rho \tilde{q} .$$

According to (4.20) and (4.21) we replace now the expression (1.14) for the production of entropy by

$$(4.23) \qquad \Gamma = \dot{H} - \oint_a \frac{h}{\vartheta} da - \int_\vartheta \rho \frac{\tilde{q}}{\vartheta} d\vartheta$$

and the Clausius-Duhem inequality (1.17) becomes

$$(4.24) \qquad \rho \vartheta \dot{\eta} - h^i_{,i} - \rho \tilde{q} + \frac{1}{\vartheta} h^i \vartheta_{,i} \geq 0 ,$$

or

$$(4.25) \qquad - \rho \dot{\psi} + \rho \dot{\varepsilon} - \rho \eta \dot{\vartheta} - h^i_{,i} - \rho \tilde{q} + \frac{1}{\vartheta} h^i \vartheta_{,i} \geq 0 ,$$

where the free energy function ψ is introduced from (1.18). Using (4.22) this inequality becomes

$$(4.26) \qquad - \rho \dot{\psi} + t^{ij} g_{il} \dot{\Phi}^l_{(\lambda)} \Phi^{(\lambda)}_j - \rho \eta \dot{\vartheta} + \frac{1}{\vartheta} h^i \vartheta_{,i} \geq 0 .$$

Chapter 5

Constitutive Relations

For independent variables we select now the elastic distorsions $\Phi^{l}_{(\lambda)}$, temperature ϑ and temperature gradients $\vartheta_{,i}$. According to the principle of equipresence we may assume that the constitutive relations are of the form

$$\psi = \psi(\Phi^{l}_{(\lambda)}, \vartheta, \vartheta_{,i}) , \tag{5.1}$$

$$h^{i} = h^{i}(\Phi^{l}_{(\lambda)}, \vartheta, \vartheta_{,i}) , \tag{5.2}$$

$$\eta = \eta(\Phi^{l}_{(\lambda)}, \vartheta, \vartheta_{,i}) , \tag{5.3}$$

$$t^{ij} = t^{ij}(\Phi^{l}_{(\lambda)}, \vartheta, \vartheta_{,i}) . \tag{5.4}$$

A set of constitutive relations is admissible (*) only if the entropy production Γ is non-negative for every process compatible with these constitutive relations. Besides the restrictions for constitutive equations which follow from the entropy inequality, some restrictions follow also from the principle of objectivity.

(*) COLEMAN and NOLL [12] .

With (5.1) the Clausius-Duhem inequality obtains the form

$$(5.5)\qquad \left(-\varrho\frac{\partial\Psi}{\partial\Phi^l_{(\lambda)}}+t^{ij}_l\Phi^{(\lambda)}_j\right)\dot{\Phi}^l_{(\lambda)}-\varrho\left(\frac{\partial\Psi}{\partial\vartheta}+\eta\right)\dot{\vartheta}-\varrho\frac{\partial\Psi}{\partial\vartheta_{,i}}\dot{\overline{\vartheta_{,i}}}+\frac{1}{\vartheta}h^i\vartheta_{,i}\geq 0\,.$$

This inequality will be satisfied for arbitrary rates $\dot{\Phi}^l_{(\lambda)}$, $\dot{\vartheta}$, $\dot{\overline{\vartheta_{,i}}}$ only if

$$(5.6)\qquad t^{ij}=\varrho g^{il}\frac{\partial\Psi}{\partial\Phi^l_{(\lambda)}}\Phi^j_{(\lambda)}\quad ,$$

$$(5.7)\qquad \eta=-\frac{\partial\Psi}{\partial\vartheta}\quad ,$$

$$(5.8)\qquad \frac{\partial\Psi}{\partial\vartheta_{,i}}=0\quad ,$$

and (5.5) reduces to the heat conduction inequality

$$(5.9)\qquad h^i\vartheta_{,i}\geq 0\,.$$

Hence from the entropy inequality it follows that Ψ cannot be a function of the temperature gradient and consequently $\vartheta_{,i}$ has to be dropped out of the equations for Ψ and η .

The principle of objectivity applied to the con stitutive relations

$$\Psi=\Psi(\Phi^l_{(\lambda)},\vartheta)\ ,$$

$$\eta = \eta(\Phi^l_{(\lambda)}, \theta) ,$$

$$t^{ij} = t^{ij}(\Phi^l_{(\lambda)}, \theta) , \tag{5.10}$$

$$h^i = h^i(\Phi^l_{(\lambda)}, \theta, \theta_{,i}) ,$$

requires that

$$\overset{*}{\Psi} = \overset{*}{\Psi}(\overset{*}{\Phi}{}^l_{(\lambda)}, \theta) = \Psi(\Phi^l_{(\lambda)}, \theta) ,$$

$$\overset{*}{\eta} = \overset{*}{\eta}(\overset{*}{\Phi}{}^l_{(\lambda)}, \theta) = \eta(\Phi^l_{(\lambda)}, \theta) ,$$

$$\overset{*}{t}{}^{ij} = \overset{*}{t}{}^{ij}(\overset{*}{\Phi}{}^l_{(\lambda)}, \theta) = t^{mn}(\Phi^l_{(\lambda)}, \theta) Q^i_{\cdot m} Q^j_{\cdot n} , \tag{5.11}$$

$$\overset{*}{h}{}^i = \overset{*}{h}{}^i(\overset{*}{\Phi}{}^l_{(\lambda)}, \theta, \overset{*}{\theta}_{,i}) = h^m(\Phi^l_{(\lambda)}, \theta, \theta_{,i}) Q^i_{\cdot m} ,$$

where

$$\overset{*}{\Phi}{}^l_{(\lambda)} = \Phi^m_{(\lambda)} Q^l_{\cdot m} ,$$

$$\overset{*}{\theta}_{,i} = \theta_{,m} Q_i^{\cdot m} . \tag{5.12}$$

To investigate the consequences of this requirement we may use the method proposed by TOUPIN (*) [44] in the theory of elastic dielectrica.

Let $Q^{l}_{\cdot m}$ be an orthogonal transformation

$$Q^{l}_{\cdot m} = \delta^{l}_{\cdot m} + \omega^{l}_{\cdot m} ,$$

where $\omega^{l}_{\cdot m}$ is an arbitrary infinitesimal rotation. From $(5.11)_1$ and $(5.12)_1$ follows that the function Ψ will be invariant under this transformation only if

$$\left(g^{il} \frac{\partial \Psi}{\partial \Phi^{l}_{(\lambda)}} \Phi^{j}_{(\lambda)} \right)_{[ij]} = 0 \tag{5.13}$$

The objectivity requirement will be fulfilled if instead of the elastic distorsions we consider as independent variables the components of the material tensor of elastic deformation,

$$\Psi = \Psi(\overset{\varepsilon}{C}_{\lambda\mu}, \vartheta) ,$$

(*) In fact the requirement that Ψ is an objective function is equivalent with the requirement that the right-hand side of (5.6) is symmetric, which is in agreement with the second law of Cauchy and which gives

$$\left(g^{il} \frac{\partial \Psi}{\partial \Phi^{l}_{(\lambda)}} \Phi^{j}_{(\lambda)} \right)_{[ij]} = 0 .$$

(see also TOUPIN [46]).

$$\eta = \eta(\overset{E}{C}_{\lambda\mu}, \vartheta) ,$$

$$t^{ij} = t^{ij}(\overset{E}{C}_{\lambda\mu}, \vartheta) , \tag{5.14}$$

$$h^{i} = h^{i}(\overset{E}{C}_{\lambda\mu}, \vartheta, \vartheta_{,\lambda}),$$

where

$$\overset{E}{C}_{\lambda\mu} = g_{lm}\Phi^{l}_{(\lambda)}\Phi^{m}_{(\mu)} \tag{5.15}$$

and

$$\vartheta_{,\lambda} = \Phi^{i}_{(\lambda)}\vartheta_{,i} . \tag{5.16}$$

Using the relations (2.18) it is possible to express the elastic distorsions in terms of the gradients of the total deformation,

$$\theta^{l}_{(\lambda)} = x^{l}_{;L}\,\theta^{L}_{(\lambda)} \tag{5.17}$$

This representation is of great importance since in this way nine unknown functions $\Phi^{l}_{(\lambda)}$ are substituted by only three unknown total deformations $x^{i} = x^{i}(\underset{\sim}{X}, t)$. With (5.17) we obtain for (5.15) and (5.16)

$$\overset{E}{C}_{\lambda\mu} = C_{LM}\,\theta^{L}_{(\lambda)}\theta^{M}_{(\mu)} , \tag{5.18}$$

(5.19) $$\theta_{,\lambda} = \theta_{,L}\theta^{L}_{(\lambda)} .$$

Here C_{LM} is the material tensor of total deformation,

(5.20) $$C_{LM} = g_{lm} x^{l}_{;L} x^{m}_{;M}$$

and $\theta_{,L} = \partial\theta/\partial X^{L}$ is the material temperature gradient.

From (5.6) and $(5.14)_1$ we obtain now for stress the expression

(5.21) $$t^{ij} = 2\varrho \frac{\partial\Psi}{\partial \overset{E}{C}_{\lambda\mu}} \Phi^{i}_{(\lambda)}\Phi^{j}_{(\mu)}$$

The constitutive equation for h^{i} may be also written in the form

(5.22) $$h^{i} = h^{i}(C_{KL}, \theta, \theta_{,L}) ,$$

which coincides with the form derived by PIPKIN and RIVLIN [31] and by GREEN and ADKINS [20] . As already mentioned in the section 1, Pipkin and Rivlin demonstrated that h^{i} vanishes when $\theta_{,L}$ vanishes. This suggests that h^{i} has the form of Fourier's law of heat conduction,

(5.23) $$\underset{\sim}{h} = \underset{\sim}{K}(\underset{\sim}{C}, \theta)\, \mathrm{grad}_{\underset{\sim}{X}}\theta ,$$

where $\mathrm{grad}_{\underset{\sim}{X}}\theta = \left\{\frac{\partial\theta}{\partial X^{L}}\right\}$. COLEMAN and NOLL [12] demonstrated that for all materials whose group of material symmetries implies the reflection $\underset{\sim}{h}$ must be an odd function of the temperature gradient.

A material is <u>isotropic</u> in a reference configuration if the constitutive relations are invariant under the full group of orthogonal transformations of material coordinates (i.e. of the coordinate system to which the reference configuration is related).

For thermoelastic stress reference configuration is the imagined non-Euclidean configuration K. Under thermal deformations each neighbourhood $N(X)$ in a body suffers a homogeneous deformation. The configuration K consists of neighbourhoods $\tilde{N}(X)$ obtained from homogeneously deformed neighbourhoods $N(X)$ in the initial configuration K_0 by local homogeneous deformations. If a body was isotropic in the initial configuration, each neighbourhood $\tilde{N}(X)$ will remain locally isotropic after a homogeneous deformation.

Considering the neighbourhoods $\tilde{N}(X)$ as locally Euclidean, for an isotropic body we require that the constitutive relations (5.14) are isotropic functions at all points $\underset{\sim}{X}$ of the body under local orthogonal transformations of coordinates u^{λ}.

If $\Psi(\overset{E}{C}_{\lambda\mu})$ is an isotropic function, then (see e.g. TRUESDELL and NOLL [45])

$$\Psi = \Psi(I^{E}_{\underset{\sim}{C}}, II^{E}_{\underset{\sim}{C}}, III^{E}_{\underset{\sim}{C}}) , \tag{5.24}$$

where $\underset{\sim}{I}_C^E$, $\underset{\sim}{II}_C^E$, $\underset{\sim}{III}_C^E$ are principal invariants of the tensor $\overset{E}{C}_{\lambda\mu}$ given by (3.30). According to (5.18) we may consider also Ψ as a function of C_{LM} and ϑ , which agrees with the general theory of thermomechanics outlined in the section 1, and which is generally accepted in the literature (cf. for instance SIGNORINI [34] , [35] , GREEN and ADKINS [20] , SEDOV [33] , PARKUS [30] , GRIOLI [23] , TRUESDELL and NOLL [49]). For the purposes of further analysis it seems more appropriate to consider the elastic and not the total deformation as an independent variable in the stress relation.

From (3.32) we see that Ψ may be considered as a function of the invariants of the tensor $\overset{E}{c}_{lm}$, and according to (3.21) and (3.22) as a function of the invariants of the tensors $\overset{E}{E}_{\lambda\mu}$ or $\overset{E}{e}_{lm}$

To write the stress relation in terms of the spatial deformation tensor we shall replace the distorsions $\Phi^l_{(\lambda)}$ in (5.6) by their reciprocals $\Phi^{(\mu)}_m$. Using (2.17) we obtain

$$t^{ij} = -\rho g^{il} \frac{\partial \Psi}{\partial \Phi^{(\mu)}_j} \Phi^{(\mu)}_l . \tag{5.25}$$

From the symmetry of the stress tensor it follows that Ψ has to satisfy the relations

$$\left(g^{il} \frac{\partial \Psi}{\partial \Phi^{(\mu)}_j} \Phi^{(\mu)}_l \right)_{[ij]} = 0 .$$

The integrals of these equations are

$$\overset{-1}{C}{}_E^{\lambda\mu} = g^{lm}\Phi_l^{(\lambda)}\Phi_m^{(\mu)} \tag{5.26}$$

and Ψ is an arbitrary function of the tensor $\overset{-1}{\underset{\sim}{C}}{}_E$.

For isotropic materials Ψ may be considered as a function of the invariants of the tensor $\overset{-1}{\underset{\sim}{C}}{}_E$, but since

$$I^E_{\overset{-1}{\underset{\sim}{C}}} = I^E_{\underset{\sim}{C}}\ , \quad II^E_{\overset{-1}{\underset{\sim}{C}}} = II^E_{\underset{\sim}{C}}\ , \quad III^E_{\overset{-1}{\underset{\sim}{C}}} = III^E_{\underset{\sim}{C}}\ , \tag{5.27}$$

for isotropic materials we shall consider Ψ as a function of the form

$$\Psi = \Psi(I^E_{\underset{\sim}{C}},\ II^E_{\underset{\sim}{C}}\ ,\ III^E_{\underset{\sim}{C}})\ . \tag{5.28}$$

According to (3.8) we have

$$a_{lm} = c_{lm} = a_{\lambda\mu}\Phi_l^{(\lambda)}\Phi_m^{(\mu)}\ ,$$

and from (5.25) we easily obtain (STOJANOVIC [36])

$$t^{ij} = -2\rho g^{il}\frac{\partial\Psi}{\partial \overset{E}{c}_{jm}}\overset{E}{c}_{ml} = \rho g^{il}\frac{\partial\Psi}{\partial \overset{E}{e}_{jm}}(g_{ml} - 2\overset{E}{e}_{ml})\ . \tag{5.29}$$

The stress relation (5.29) will be identically symmetric in i and j only if Ψ is of the form (5.28), i.e. if the material is isotropic.

The principal invariants of the spatial tensor of elastic strain $\overset{E}{\underset{\sim}{e}}$ are

$$\mathrm{I}^{E}_{\underset{\sim}{e}} = \delta^{i}_{l}\overset{E}{e}{}^{l}_{i} ,$$

(5.30)
$$\mathrm{II}^{E}_{\underset{\sim}{e}} = \frac{1}{2!}\delta^{ij}_{lm}\overset{E}{e}{}^{l}_{i}\overset{E}{e}{}^{m}_{j} ,$$

$$\mathrm{III}^{E}_{\underset{\sim}{e}} = \frac{1}{3!}\delta^{ijk}_{lmn}\overset{E}{e}{}^{l}_{i}\overset{E}{e}{}^{m}_{j}\overset{E}{e}{}^{n}_{k} .$$

Assuming now Ψ in the form

(5.31)
$$\Psi = \Psi(\mathrm{I}^{E}_{\underset{\sim}{e}}, \mathrm{II}^{E}_{\underset{\sim}{e}}, \mathrm{III}^{E}_{\underset{\sim}{e}}) ,$$

using (5.30) and the Cayley-Hamilton theorem

(5.32)
$$\overset{3E}{\underset{\sim}{e}} = \mathrm{III}^{E}_{\underset{\sim}{e}}\underset{\sim}{1} + \mathrm{II}^{E}_{\underset{\sim}{e}}\overset{E}{\underset{\sim}{e}} + \mathrm{I}^{E}_{\underset{\sim}{e}}\overset{2E}{\underset{\sim}{e}} ,$$

where $\underset{\sim}{1}$ is the unit tensor and

(5.33)
$$\overset{2E}{e} = \{e^{i}_{m}e^{m}_{j}\} , \qquad \overset{3E}{e} = \{e^{i}_{m}e^{m}_{n}e^{n}_{j}\} ,$$

from (5.29) we obtain

(5.34)
$$t^{ij} = G_0 g^{ij} + G_1\overset{E}{e}{}^{ij} + G_2\overset{E}{e}{}^{i}_{m}e^{mj} .$$

Here we introduce the notation

$$G_0 = \rho\left[\frac{\partial\Psi}{\partial\mathrm{I}^{E}_{\underset{\sim}{e}}} + \mathrm{I}^{E}_{\underset{\sim}{e}}\frac{\partial\Psi}{\partial\mathrm{II}^{E}_{\underset{\sim}{e}}} + (\mathrm{II}^{E}_{\underset{\sim}{e}} - 2\mathrm{III}^{E}_{\underset{\sim}{e}})\frac{\partial\Psi}{\partial\mathrm{III}^{E}_{\underset{\sim}{e}}}\right] ,$$

$$G_1 = -\rho\left[2\frac{\partial\Psi}{\partial I^E_{\underset{\sim}{e}}} + (1 + 2I^E_{e})\frac{\partial\Psi}{\partial II^E_{e}} + I^E_{e}\frac{\partial\Psi}{\partial III^E_{e}}\right],$$

$$G_2 = \rho\left[2\frac{\partial\Psi}{\partial II^E_{e}} + \frac{\partial\Psi}{\partial III^E_{e}}\right]. \tag{5.35}$$

The elastic strains $\overset{E}{\underset{\sim}{E}}$ and $\overset{E}{\underset{\sim}{e}}$ do not satisfy the compatibility conditions and in order to solve any problem of thermoelasticity we have to introduce compatible strains $\underset{\sim}{E}$ or $\underset{\sim}{e}$ which correspond to the total deformation. This can be achieved using (3.25) or (3.26) and if the thermal strain is known, the problem of thermoelasticity is completely determined, i.e. the number of unknown functions is equal to the number of available equations.

However, using (5.18) the internal energy function may be also assumed in the form

$$\Psi = \Psi(\underset{\sim}{C}, \theta, \underset{\sim}{X}) \tag{5.36}$$

or

$$\Psi = \Psi(\underset{\sim}{E}, \theta, \underset{\sim}{X}) \tag{5.37}$$

and the stress relation becomes

$$t^{ij} = \rho\frac{\partial\Psi}{\partial E_{KL}}x^i_{;K}x^j_{;L}\ . \tag{5.38}$$

If the second Piola-Kirchhoff tensor T^{KL} is introduced instead of the Cauchy stress t^{ij},

$$T^{KL} = \frac{\rho_0}{\rho} X^K_{;i} X^L_{;j} t^{ij} , \tag{5.39}$$

the stress relation reduces to

$$T^{KL} = \rho_0 \frac{\partial \Psi}{\partial E_{KL}} , \tag{5.40}$$

and all quantities are referred to the initial configuration K_0.

CHADWICK and SEET [9] use (5.40) to define the <u>material constants</u> in the reference configuration K_0. The <u>isothermal elasticities</u> of order n are defined by

$$C^{K_1L_1 \ldots K_nL_n} = \rho_0 \left(\frac{\partial^n \Psi}{\partial E_{K_1L_1} \ldots \partial E_{K_nL_n}} \right)_0 , \quad n = 2, 3 \ldots , \tag{5.41}$$

and the <u>temperature coefficients</u> of these elasticities are

$$\underset{T}{C}^{K_1L_1 \ldots K_nL_n} = \rho_0 \left(\frac{\partial^{n+1} \Psi}{\partial T \partial E_{K_1L_1} \ldots \partial E_{K_nL_n}} \right)_0 . \tag{5.42}$$

The <u>temperature coefficients of stress</u> are

$$\beta^{KL} = -\left(\frac{\partial T^{KL}}{\partial T} \right) = -\rho_0 \left(\frac{\partial^2 \Psi}{\partial T \partial E_{KL}} \right)_0 ,$$

$$\underset{1}{\beta}^{KL} = -\left(\frac{\partial^2 T^{KL}}{\partial T^2} \right) = -\rho_0 \left(\frac{\partial^3 \Psi}{\partial T^2 \partial E_{KL}} \right)_0 , \tag{5.43}$$

. . . .

The entropy relation (5.7) may be written in the form

$$\eta = -\frac{\partial \Psi}{\partial T} , \qquad T = \theta - \theta_0 \tag{5.44}$$

and for sufficiently small increments of temperature we may also write

$$\eta = -\left(\frac{\partial \Psi}{\partial T}\right)_0 - \left(\frac{\partial^2 \Psi}{\partial T_2}\right)_0 T + \dots \tag{5.45}$$

where $\left(\frac{\partial \Psi}{\partial T}\right)_0 = -\eta_0$ is the entropy of the initial configuration,

$$\eta = \eta_0 + \left(\frac{\partial \eta}{\partial T}\right)_0 T + \dots. \tag{5.46}$$

The coefficient

$$C = \theta_0 \left(\frac{\partial \eta}{\partial T}\right)_0 = -\theta_0 \left(\frac{\partial^2 \Psi}{\partial T^2}\right)_0 , \tag{5.47}$$

represents the <u>specific heat</u> at constant deformation, and

$$C_T = \left[\frac{\partial}{\partial \theta}\left(\theta \frac{\partial \eta}{\partial \theta}\right)\right]_0 = \frac{C}{\theta_0} - \theta_0 \left(\frac{\partial^3 \Psi}{\partial T_3}\right)_0 , \tag{5.48}$$

represents the <u>temperature coefficient of the specific heat.</u>

Chapter 6

Approximation for Isotropic Materials

After the analysis of the preceding section the constitutive equations of thermoelasticity may be written as follows:

(6.1) $$\Psi = \Psi(\overset{E}{E}_{\lambda\mu}, \vartheta) ,$$

(6.2) $$\eta = - \frac{\partial \Psi}{\partial \vartheta} ,$$

(6.3) $$h^i = h^i(\overset{E}{E}_{\lambda\mu}, \vartheta, \vartheta_{,L}) ,$$

(6.4) $$t^{ij} = \rho \frac{\partial \Psi}{\partial \overset{E}{E}_{\lambda\mu}} \Phi^i_{(\lambda)} \Phi^j_{(\mu)} ,$$

where the tensor $\overset{E}{E}_{\lambda\mu}$ of the elastic strain is referred to the non-Euclidean configuration K. The elastic strain tensor may be substituted by the tensor E_{LM} of the total deformation, re ferred to the initial configuration K_0, where

(6.5) $$\begin{aligned} \overset{E}{E}_{\lambda\mu} &= E_{LM}\theta^L_{(\lambda)}\theta^M_{(\mu)} - \overset{T}{e}_{\lambda\mu} \\ &= (E_{LM} - \overset{T}{E}_{LM})\theta^L_{(\lambda)}\theta^M_{(\mu)} = \overset{E}{E}_{LM}\theta^L_{(\lambda)}\theta^M_{(\mu)} . \end{aligned}$$

In order to apply constitutive equations to any problem of thermoelasticity we have to consider their approximate forms. The usual approximation for the heat conduction law is

$$h_K = K_K^L(\underset{\sim}{C}, \vartheta, \vartheta_{,M})\vartheta_{,L} \tag{6.6}$$

PIPKIN and RIVLIN [31] demonstrated that for isotropic materials this law may be written in the form

$$h_k = K_k^l(\underset{\sim}{c}, \vartheta, \vartheta_{,m})\vartheta_{,l} , \tag{6.7}$$

where

$$K_k^l = h_0\delta_k^l + h_1 c_k^l + h_2 c_k^t c_t^l \tag{6.8}$$

and h_r are functions of ϑ and of the following six invariants;

$$I_{\underset{\sim}{c}},\ II_{\underset{\sim}{c}},\ III_{\underset{\sim}{c}},\ (\mathrm{grad}\,\vartheta)^2,\ \ c^{kl}\vartheta_{,k}\vartheta_{,l}\ ,\ \ c_m^l c^{mk}\vartheta_{,k}\vartheta_{,l} \tag{6.9}$$

An approximation of any desired degree of accuracy may be obtained directly from these relations.

CHADWICK and SEET [9] in the second-order theory, but for any type of material symmetries approximated (6.6) by the relation

(6.10) $$h^K = (k^{KL} + m^{KLMN}E_{MN} + \frac{1}{2}k^{KLM}\vartheta_{,M} + k_T^{KL}\vartheta)\vartheta_{,L} + O(\nu^3) \quad ,$$

where $O(\nu^3)$ stands for the terms of the third and higher order.

The free-energy function Ψ may be expanded into a series in terms of the elastic strain tensor in the vicinity of the reference configuration K_0. The energy per unit volume of the body will be

(6.11) $$\rho_\vartheta \Psi = J_\vartheta \rho_0 (\Psi_\vartheta + \sum_{i=1}^{\infty} A^{\lambda_1\mu_1 \dots \lambda_i\mu_i} \overset{E}{E}_{\lambda_1\mu_1} \dots \overset{E}{E}_{\lambda_i\mu_i}) \quad ,$$

where ρ_0 is the density of matter in the reference configuration, Ψ_ϑ is the energy stored by the thermal deformation and it is assumed that the anisotropic tensors $A^{\cdots}$ are functions of temperature . $A^{\cdots}$ are isothermal tensor-valued coefficients and may be determined from a static experiment performed over a body at a uniform temperature ϑ . In (6.11) we have put

(6.12) $$J_\vartheta = (1 + 2\,\mathrm{I}_{\underset{\sim}{E}}^{E} + 4\,\mathrm{II}_{\underset{\sim}{E}}^{E} + 8\,\mathrm{III}_{\underset{\sim}{E}}^{E})^{-1/2} .$$

To calculate the coefficients $A^{\cdots}$ let us consider an infinitesimal displacement ω^k superposed upon a homogeneous thermal deformation. If $\underset{\sim}{E}$ is the total strain, $\underset{\sim}{F}$ the strain of the superposed deformation with respect to the previously homogeneously deformed body, the energy per unit volume of the body may be calculated in two ways, for the total deformation and for the superposed deformation, but the total amount of the energy remains unchanged,

$$\rho_{\underset{\sim}{E}} \Psi(\underset{\sim}{E}) = \rho_{\underset{\sim}{F}} \Psi(\underset{\sim}{F}) \ . \tag{6.13}$$

According to the conservation of mass we have

$$\rho \, d\vartheta = \rho_{\underset{\sim}{F}} \, d\vartheta_{\underset{\sim}{F}} = \rho_0 \, dv \tag{6.14}$$

and since

$$\begin{aligned} \rho_{\underset{\sim}{E}} &= \rho_0 (1 + 2\,I_{\underset{\sim}{E}} + 4\,II_{\underset{\sim}{E}} + 8\,III_{\underset{\sim}{E}})^{-\frac{1}{2}} = \rho_0 \, J_{\underset{\sim}{E}} \\ \rho_{\underset{\sim}{F}} &= \rho_0 (1 + 2\,I_{\underset{\sim}{F}} + 4\,II_{\underset{\sim}{F}} + 8\,III_{\underset{\sim}{F}})^{-\frac{1}{2}} = \rho_0 \, J_{\underset{\sim}{F}} \end{aligned} \tag{6.15}$$

it follows that

$$J_{\underset{\sim}{E}} \Psi(\underset{\sim}{E}) = J_{\underset{\sim}{F}} \Psi(\underset{\sim}{F}) \ . \tag{6.16}$$

From (2.25) for a homogeneous thermal deformation we may write

$$u^\lambda = (\delta^\lambda_K + \bar{\alpha}^\lambda_{\cdot K}) X^K \ , \tag{6.17}$$

where X^K, u^λ and x^k are Cartesian coordinates. The total deformation will be

$$x^k = \delta^k_\lambda u^\lambda + \omega^k \ . \tag{6.18}$$

The displacement ω^k is superposed upon the homogeneously deformed body and therefore we have to consider the displacement gradients $\omega^k_{,\lambda}$ with respect to the intermediate coordinates u^λ . The total deformation gradients are

$$x^k_{;K} = \delta^k_K + \bar{\alpha}^k_{\cdot K} + \omega^k_{,\lambda}(\delta^\lambda_K + \bar{\alpha}^\lambda_{\cdot K}) \tag{6.19}$$

and the total strain tensor will be

$$E_{LM} = T_{LM} + P^{\lambda\mu}_{LM} F_{\lambda\mu} , \tag{6.20}$$

where

$$2T_{LM} = \bar{\alpha}_{LM} + \bar{\alpha}_{ML} + \bar{\alpha}_{\lambda L}\bar{\alpha}^\lambda_{\cdot M} ,$$

$$P^{\lambda\mu}_{LM} = \delta^\lambda_L \delta^\mu_M + \delta^\lambda_L \bar{\alpha}^\mu_{\cdot M} + \delta^\mu_M \bar{\alpha}^\lambda_{\cdot L} + \bar{\alpha}^\lambda_{\cdot L}\bar{\alpha}^\mu_{\cdot M} , \tag{6.21}$$

$$F_{\lambda\mu} = \omega_{(\lambda,\mu)} + \frac{1}{2}\omega_{\nu,\lambda}\omega^\nu_{,\mu} .$$

Introducing E_{LM} from (6.20) into the series

$$\rho\Psi(\underset{\sim}{E}) = \rho_0 J_{\underset{\sim}{E}} \sum_{i=1}^{\infty} A^{L_1M_1\ldots L_iM_i} E_{L_1M_1}\ldots E_{L_iM_i} \tag{6.22}$$

from (6.16) we obtain

$$\Psi(\underset{\sim}{F}) = \frac{J_{\underset{\sim}{E}}}{J_{\underset{\sim}{F}}}\left(\alpha + \sum_{i=1}^{\infty} A_\theta^{\lambda_1\mu_1\ldots\lambda_i\mu_i} F_{\lambda_1\mu_1}\ldots F_{\lambda_i\mu_i}\right) \tag{6.23}$$

where $J_{\underset{\sim}{E}}\alpha/J_{\underset{\sim}{F}}$ is the specific energy of the body stored by heating, and $\frac{J_{\underset{\sim}{E}}}{J_{\underset{\sim}{F}}} A_\theta^{\lambda_1\mu_1\ldots\lambda_i\mu_i}$ are the elasticity tensors for the body at a temperature θ . This procedure was applied by BRILLOUIN [3] , [7] for the determination of the coefficients of elasticity for isotropic bodies at different temperatures. (cf. also TRUESDELL [47] and GREEN [21]). SIGNORINI [25] proved that in isotropic bodies heating induces a uniform expansion and that the isotropy is not violated by heating. Therefore for a body heated from a reference temperature ϑ_0 to a uniform temperature $\vartheta = \vartheta_0 + T$ we have in (6.17) $\bar{\alpha}^{\lambda}_{\cdot K} = \bar{\alpha}\,\delta^{\lambda}_{K}$ and writing $1 + \bar{\alpha} = K$ we obtain from (6.21).

$$2T_{LM} = (2\alpha + \alpha^2)\delta_{LM} = (K^2 - 1)\delta_{LM}\,, \quad P^{\lambda\mu}_{LM} = K^2 \delta^{\lambda}_{L}\delta^{\mu}_{M}\,. \tag{6.24}$$

The tensor of total strain E_{LM} becomes

$$E_{LM} = \frac{1}{2}(K^2 - 1)\delta_{LM} + K^2 F_{LM}\,. \tag{6.25}$$

The invariants of the tensor E_{LM} are

$$\begin{aligned} I_{\underset{\sim}{E}} &= \frac{3}{2}(K^2 - 1) + K^2 I_{\underset{\sim}{F}}\,, \quad II_{\underset{\sim}{E}} = \frac{3}{4}(K^2-1)^2 + K^2(K^2-1)I_{\underset{\sim}{F}} + K^4 II_{\underset{\sim}{F}}\,, \\ III_{\underset{\sim}{E}} &= \frac{1}{8}(K^2 - 1) + \frac{1}{4}K^2(K^2-1)^2 I_{\underset{\sim}{F}} + \frac{1}{2}K^4(K^2-1) II_{\underset{\sim}{F}} + K^6 III_{\underset{\sim}{F}}\,, \end{aligned} \tag{6.26}$$

and we easily find

$$J_{\underset{\sim}{E}} = K^3 J_{\underset{\sim}{F}}\,. \tag{6.27}$$

Writing for the coefficients of elasticity at the reference temperature ϑ_0, $\lambda, \mu, l, m, n, \ldots$, the relation (6.13) for isotropic materials may be written

$$(6.28)\quad K^3\Psi(\underset{\sim}{F}) = \Psi(\underset{\sim}{E}) = \frac{\lambda+2\mu}{2} I_{\underset{\sim}{E}}^2 - 2\mu II_{\underset{\sim}{E}} + l I_{\underset{\sim}{E}}^3 + m I_{\underset{\sim}{E}} II_{\underset{\sim}{E}} + n III_{\underset{\sim}{E}} + \dots,$$

where $\lambda, \mu, l, m, n, \dots,$ are constants. But we may write also

$$(6.29)\quad \Psi(\underset{\sim}{F}) = \Psi_0 + p I_{\underset{\sim}{F}} + \frac{\lambda^* + 2\mu^*}{2} I_{\underset{\sim}{F}}^2 - 2\mu^* II_{\underset{\sim}{F}} + l^* I_{\underset{\sim}{F}}^3 + m^* I_{\underset{\sim}{F}} II_{\underset{\sim}{F}} + n^* III_{\underset{\sim}{F}} + \dots,$$

where Ψ_0 is energy stored by heating from ϑ_0 to ϑ, p is the uniform hydrostatic tension at the temperature ϑ, λ^* and μ^* are Lamé constants and l^*, m^* and n^* are the third-order elastic constants at ϑ. From (6.26), (6.28) and (6.29) follow the following relations:

$$K^3\Psi_0 = \frac{1}{8}(K^2-1)^2[3(3\lambda + 2\mu) + (K^2-1)(2 + l + 18m + n) + \dots],$$

$$Kp = (K^2-1)^2\Big[\frac{3\lambda + 2\mu}{2} + (K^2-1)\frac{2 + l + 9m + n}{4} + \dots\Big],$$

$$(6.30)\quad \lambda^* = K\Big[\lambda + (K^2-1)\frac{18l + 7m + n}{2} + \dots\Big],$$

$$\mu^* = K\Big[\mu + (K^2-1)\frac{3m + 2n}{4} + \dots\Big],$$

$$l^* = K^3(l + \dots), \quad m^* = K^3(m + \dots), \quad n^* = K^3(n + \dots),$$

In a non-uniformly heated body the elasticity coefficients are functions of position.

If the thermal expansion in a homogeneous isotropic body is described with sufficient accuracy by

$$K = 1 + \alpha T(X) \quad , \qquad \alpha = \text{const.} \tag{6.31}$$

from (3.15) we find for the thermal strain tensor

$$\overset{T}{E}_{LM} = \frac{1}{2}(K^2 - 1)G_{LM} \tag{6.32}$$

and the thermal distorsions are given by (3.11),

$$\theta_L^{(\lambda)} = K\delta_L^\lambda \tag{6.33}$$

Let us assume that the elastic and total strains and αT are of the same order of magnitude ν, $|\nu| < 1$.

According to (5.31), for isotropic bodies we may expand the function Ψ in the vicinity of the non-Euclidean configuration K into a series

$$\Psi = \frac{\bar{\lambda} + 2\bar{\mu}}{2}(I_{\underset{\sim}{e}}^E)^2 - 2\bar{\mu}II_{\underset{\sim}{e}}^E + \bar{l}(I_{\underset{\sim}{e}}^E)^3 + \bar{m}I_{\underset{\sim}{e}}^E II_{\underset{\sim}{e}}^E + \bar{n}III_{\underset{\sim}{e}}^E + \ldots \tag{6.34}$$

Using (3.24) we find the relations between the invariants of the tensors $\overset{E}{e}_{lm}$ and $\overset{E}{E}_{LM}$,

$$I_{\underset{\sim}{e}}^E = \overset{-1}{C}{}^{AB}\overset{E}{E}_{AB} \ ,$$

$$II_{\underset{\sim}{e}}^E = \frac{1}{2!}\overset{-1}{C}{}^{AL}\overset{-1}{C}{}^{BM}\delta_{AB}^{PQ}\overset{E}{E}_{PL}\overset{E}{E}_{QM} \ , \tag{6.35}$$

$$III_{\underset{\sim}{e}}^E = \frac{1}{3!}\overset{-1}{C}{}^{AL}\overset{-1}{C}{}^{BM}\overset{-1}{C}{}^{CN}\delta_{ABC}^{PQR}\overset{E}{E}_{PL}\overset{E}{E}_{QM}\overset{E}{E}_{RN} \ ,$$

where

$$(6.36) \qquad \overset{-1}{C}{}^{AB} = g^{ij} X^{A}_{;i} X^{B}_{;j}$$

or

$$(6.37) \qquad \overset{-1}{\underset{\sim}{C}} = (\underset{\sim}{G} + 2\underset{\sim}{E})^{-1} ,$$

which gives

$$(6.38) \qquad \overset{-1}{C}{}^{AB} = G^{AB} - 2E^{AB} + 4E^{A}_{S} E^{SB} + O(E^{3}) .$$

Using the relation (3.5) we obtain from (6.32)

$$(6.39) \qquad E^{AB} = \underset{E}{E}{}^{AB} + \frac{1}{2}(K^{2} - 1)G^{AB} ,$$

and if we write $K - 1 = \Omega = \alpha T$, $\overset{-1}{C}{}^{AB}$ reduces to

$$(6.40) \qquad \overset{-1}{C}{}^{AB} = (1 - 2\Omega + 3\Omega^{2})G^{AB} - 2(1 - 4\Omega)\underset{E}{E}{}^{AB} + 4\underset{E}{E}{}^{A}_{S}\underset{E}{E}{}^{SB} + O(E^{3}) .$$

From (6.35) now we find

$$(6.41) \qquad \begin{aligned} \mathrm{I}^{E}_{\underset{\sim}{e}} &= (1 - 2\Omega)\mathrm{I}^{E}_{\underset{\sim}{E}} - 2(\mathrm{I}^{E}_{\underset{\sim}{E}})^{2} + 4\mathrm{II}^{E}_{\underset{\sim}{E}} + O(\nu^{3}) , \\ \mathrm{II}^{E}_{\underset{\sim}{e}} &= (1 - 4\Omega)\mathrm{II}^{E}_{\underset{\sim}{E}} - 2\mathrm{I}^{E}_{\underset{\sim}{E}}\mathrm{II}^{E}_{\underset{\sim}{E}} + 6\mathrm{III}^{E}_{\underset{\sim}{E}} + O(\nu^{4}) , \\ \mathrm{III}^{E}_{\underset{\sim}{e}} &= \mathrm{III}^{E}_{\underset{\sim}{E}} + O(\nu^{4}) . \end{aligned}$$

Introducing the invariants of the tensor $\overset{E}{\underset{\sim}{e}}$ from (6.41) into (6.34) and comparing it with

$$\Psi = \frac{\lambda^* + 2\mu^*}{2}(I_{\underset{\sim}{E}}^E)^2 - 2\overset{*}{\mu} II_{\underset{\sim}{E}}^E + l^*(I_{\underset{\sim}{E}}^E)^3 + \overset{*}{m} I_{\underset{\sim}{E}}^E II_{\underset{\sim}{E}}^E + \overset{*}{n} III_{\underset{\sim}{E}}^E + \dots \qquad (6.42)$$

we find the relations (*) between the coefficients $\lambda^*, \mu^*, \dots$ and $\bar{\lambda}, \bar{\mu}, \dots$

$$\bar{\lambda} = \lambda^*(1 + 4\Omega) + \dots, \qquad \bar{\mu} = \overset{*}{\mu}(1 + 4\Omega) + \dots,$$

$$\bar{l} = l^* + 2(\lambda^* + 2\overset{*}{\mu})(1 + 4\Omega) + \dots, \quad \bar{m} = \overset{*}{m} - 4(\lambda^* + 3\overset{*}{\mu})(1 + 4\Omega) + \dots,$$

$$\bar{n} = \overset{*}{n} + 12\overset{*}{\mu} + \dots \qquad (6.43)$$

From (6.43) and (6.30) we obtain

$$\bar{\lambda} = \lambda + \Omega[5\lambda + 18l + 7m + n] + O(\nu^2),$$

$$\bar{\mu} = \mu + \Omega\left[5\mu + \frac{3m + n}{2}\right] + O(\nu^2), \qquad (6.44)$$

$$\bar{l} = l + 2(\lambda + 2\mu) + O(\nu), \quad \bar{m} = m - 4(\lambda + 3\mu) + O(\nu), \quad \bar{n} = n + 12\mu + O(\nu).$$

Here λ, μ, l, m, n are the elastic coefficients measured in the initial configuration K_0 at the temperature ϑ_0.

(*) For $\Omega = 0$ these relations reduce to the relations among isothermal coefficients in the spatial and material description of the energy function (cf. TRUESDELL [47], [43], where it seems that in the approximation of $I_{\underset{\sim}{e}}^2$ the term $I_{\underset{\sim}{E}}^3$ was overlooked).

The expression (6.34) for the free-energy function Ψ may be approximated by the series

$$\Psi = \frac{1}{2}(\lambda^* + 2\mu^*)(1+4\Omega)(I^E_{\underset{\sim}{e}})^2 - 2\mu^*(1+4\Omega)II^E_{\underset{\sim}{e}} + (l^* + 4\mu^* + 2\lambda^*)II^E_{\underset{\sim}{e}} + \tag{6.45}$$

$$+ (m^* - 4\lambda^* - 12\mu^*)I^E_{\underset{\sim}{e}}II^E_{\underset{\sim}{e}} + (n^* + 12\mu^*)III^E_{\underset{\sim}{e}} + O(\nu^4)$$

From (5.35) and (6.45) we easily find

$$G_0 = \rho\{[\lambda + (5\lambda + 18l + 7m + n)\Omega]I^E_{\underset{\sim}{e}} + (2\lambda + 3l + m)(I^E_{\underset{\sim}{e}})^2 +$$

$$+ (-4\lambda + m + n)II^E_{\underset{\sim}{e}}\} + O(\nu^2) \ ,$$

(6.46)

$$G_1 = \rho[2\mu - (-2\lambda + m + n)I^E_{\underset{\sim}{e}} - 8\mu\Omega] + O(\nu^2) \ ,$$

$$G_2 = \rho(8\mu + n) + O(\nu) \ ,$$

where

$$\rho = \rho_0(1 + I_e) + O(\nu^2) \tag{6.47}$$

From (3.15) and (3.23) we find

$$\overset{T}{e}_{lm} = \frac{\Omega}{2}(2 + \Omega)c_{lm} = \frac{\Omega}{2}(2 + \Omega)(g_{lm} - 2e_{lm}) \tag{6.48}$$

and

$$\overset{E}{e}_{lm} = e_{lm} - \overset{T}{e}_{lm} = (1 + \Omega)^2 e_{lm} - \frac{\Omega}{2}(2 + \Omega)g_{lm} \ . \tag{6.49}$$

The principal invariants of the tensor $\overset{E}{\underset{\sim}{e}}$ will be now

$$I_{\underset{\sim}{e}}^{E} = I_{\underset{\sim}{e}} - 3\Omega + 2\Omega I_{\underset{\sim}{e}} - \frac{3}{2}\Omega^2 + O(\nu^3) ,$$

$$II_{\underset{\sim}{e}}^{E} = II_{\underset{\sim}{e}} - 2\Omega I_{\underset{\sim}{e}} + 3\Omega^2 + O(\nu^3) , \qquad (6.50)$$

and for the coefficients G_r in (6.46) we obtain

$$G_0 = \rho_0\Big[\lambda I_{\underset{\sim}{e}} - 3\lambda\Omega + (3\lambda + 3l + m)I_{\underset{\sim}{e}}^2 + (-4\lambda + m + n)II_{\underset{\sim}{e}} +$$

$$- (m + n)\Omega I_{\underset{\sim}{e}} - 3\Big(\frac{7}{2}\lambda + 9l + 3m\Big)\Omega^2\Big] + O(\nu^3) ,$$

$$G_1 = \rho_0[2\mu + (2\lambda + 2\mu - m - n)I_{\underset{\sim}{e}} + (-6\lambda + 10\mu + 6m + 4n)\Omega] + O(\nu^2) ,$$

$$G_2 = \rho_0(8\mu + n) . \qquad (6.51)$$

The substitution of $\overset{E}{\underset{\sim}{e}}$ from (6.49) into the stress relation (5.34) yields

$$\underset{\sim}{t} = (G_0 - \Omega G_1 - \frac{1}{2}\Omega^2 G_1 + \Omega^2 G_2)\underset{\sim}{1} +$$

$$+ (G_1 + 2\Omega G_1 - 2\Omega G_2)\underset{\sim}{e} + G_2 \underset{\sim}{e}^2 , \qquad (6.52)$$

or

$$\underset{\sim}{t} = \rho_0[\lambda I_{\underset{\sim}{e}} - (3\lambda + 2\mu)\Omega + (3\lambda + 3l + m)I_{\underset{\sim}{e}}^2 + (-4\lambda + m + n)II_{\underset{\sim}{e}} +$$

$$-2(\lambda+\mu)\Omega I_{\underset{\sim}{e}} - 3\left(\frac{3}{2}\lambda+\mu+9l+5m+n\right)\Omega^2\Big]\underset{\sim}{1} +$$

$$+\rho_0[2\mu+(2\lambda+2\mu-m-n)I_{\underset{\sim}{e}}+2(-3\lambda-\mu+3m+n)\Omega]\underset{\sim}{e} +$$

$$+\rho_0(8\mu+n)\underset{\sim}{e}^2 + O(\nu^3)\,. \tag{6.53}$$

In the same form this relation was derived in 1969 by STOJANOVIC (*) [36] . With the different notation an equation equivalent to (6.51) was derived in 1971 by CHADWICK and SEET [9] . They considered the power-series approximation in the vicinity of $E = 0$ and $T = 0$ for the free energy function Ψ

$$\Psi = \Psi(\underset{\sim}{E}, \theta)\,, \tag{6.54}$$

which corresponds to $(5.14)_1$ and (5.18), and derived the second-order-approximation both for isotropic and trasversely isotropic thermoelastic materials.

Besides the Lamé coefficients λ and μ and the third order coefficients $\bar{l}$, $\bar{m}$, $\bar{n}$ given by (6.44), let us introduce the bulk modulus $\varkappa$ and the following notation

(*) In the paper [36] an error in coefficients appeared, since in the relations corresponding to (6.13) here, the equality was considered for energies per unit mass, and not per unit volume of the body.

$$\varkappa = \frac{1}{3}(3\lambda + 2\mu),$$

$$\tau_1 = -\frac{15}{2}\lambda - 7\mu + 3m + n\ , \quad \tau_3 = 3\lambda - 8\mu - 27l - 15m - 3n,$$

$$\tau_2 = 12\lambda - 22\mu - 6m - 2n\ , \quad \tau_4 = -\frac{9}{2}\lambda + 3\mu + 54l + 27m + 5n. \tag{6.55}$$

The free energy function (6.45) may be expressed in terms of the total strain $\mathbf{e}$ and the thermal dilatation Ω .

$$\Psi = \Psi_{\underset{\sim}{e}} - 3\varkappa\Omega I_{\underset{\sim}{e}} + \frac{9}{2}\varkappa\Omega^2 + \tau_1\Omega I^2_{\underset{\sim}{e}} + \tau_2\Omega II_{\underset{\sim}{e}} + $$
$$+ \tau_3\Omega^2 I_{\underset{\sim}{e}} + \tau_4\Omega^3 + O(\nu^4). \tag{6.56}$$

Here $\Psi_{\underset{\sim}{e}}$ is the part which depends only on the strain $\underset{\sim}{e}$.

To complete the approximation of the constitutive relations we have to consider also the entropy relations. The approximative relation for the specific entropy follows from (5.7) and (6.56)

$$\eta = -\frac{\partial\Psi}{\partial T} = (-3\varkappa I_{\underset{\sim}{e}} + \tau_1 I^2_{\underset{\sim}{e}} + \tau_2 II_{\underset{\sim}{e}})\alpha + [9\varkappa + 2\tau_3 I_{\underset{\sim}{e}} + 3\tau_4\Omega]\alpha\Omega + O(\nu^3). \tag{6.57}$$

Second-order approximation of the constitutive equations, derived by a direct generalization of the equations of the linear theory, were applied by HERRMANN [25] to the study of some second-order thermoelastic effects under simple shear. CHAUDHRY [10] extended to thermoelasticity the method of successive approximations developed by ADKINS, GREEN and SHIELD [1] for isothermal problems and calculated the additional stresses and displacements due to temperature for an infinite elastic body subjected to a non-uniform temperature distribution under uniform tension in one direction at infinity. The body had a circular hole.

It is not evident from these examples if there are any qualitative effects, besides the quantitative connections of the results obtained in the linear theory.

Chapter 7

Thermoelastic Bodies with Non-Symmetric Tensor

There are two kinds of theories of elastic materials for which the stress tensor is not symmetric. One kind of the theories treats the materials with microstructure, which reduce to the continua with directors, and the other kind is the theory in which the higher order deformation gradients are considered.

Without entering into details of the theories, we mention here only the basic features. In the theory of the generalized Cosserat continua at each point of a body there is assigned a number of vectors - the directors $\underset{\sim}{d}_{(1)}, \ldots, d_{(n)}$ The laws of motion for such a body are (cf. STOJANOVIC [36], [39]).

$$\rho \dot{v}^{i} = t^{ij}_{,j} + \rho f^{i} , \tag{7.1}$$

$$\rho i^{\lambda\mu} \ddot{d}^{i}_{(\mu)} = h^{(\lambda)ij}_{,j} + \rho k^{(\lambda)i} , \tag{7.2}$$

$$t^{[ij]} = d^{[i}_{(\lambda),k} h^{(\lambda)j]k} , \tag{7.3}$$

where $i^{\lambda\mu}$ are certain coefficients of microinertia, $h^{(\lambda)ij}$ are the director stresses, $k^{(\lambda)i}$ are the director forces and $t^{[ij]} = \frac{1}{2}(t^{ij} - t^{ji})$ is the antisymmetric part of the stress tensor. From (7.2) and (7.3) follows

(7.4) $$\rho\dot{\sigma}^{ij} = m^{ijk}_{\ ,k} + \rho k^{ij} - t^{[ij]} ,$$

where

(7.5) $$\sigma^{ij} = i^{\lambda\mu} d^{[i}_{(\lambda)} \dot{d}^{j]}_{(\mu)}$$

is the spin per unit mass of the body, and

$$k^{ij} = d^{[i}_{(\lambda)} k^{(\lambda)j]}$$

are certain body couples, and

$$m^{ijk} = d^{[i}_{(\lambda)} h^{(\lambda)j]k} = - m^{jik}$$

is the couple-stress tensor.

The first law of thermodynamics yields the reduced energy equation

(7.6) $$\rho\dot{\varepsilon} = t^{ij} v_{(i,j)} + h^{(\lambda)ij} \dot{d}_{(\lambda)i,j} + h^{j}_{,j} + \rho q$$

and it is natural to assume that the internal energy function is of the form

(7.7) $$\varepsilon = \varepsilon(x^{k}_{;K}, d^{k}_{\cdot(\lambda)}, d^{k}_{\cdot(\lambda);K}, \eta, X)$$

(WOZNIAK [52] , [53] , [54] , GROOT [24]). The Clausius - Duhem inequality (1.17) when combined with (7.7) gives

$$-\rho\dot{\varepsilon} + t^{ij}v_{(i,j)} + h^{(\lambda)ij}\dot{d}_{(\lambda)i,j} + \rho\theta\dot{\eta} + \frac{1}{\theta}h^{i}\theta_{,i} \geq 0. \qquad (7.8)$$

In thermoelasticity the deformation gradients corresponding to the thermal strain do not exist and we have to consider the internal function of the form

$$\varepsilon = \varepsilon(\Phi^{k}_{(\lambda)}, d^{k}_{(\lambda);K}, \eta, X), \qquad (7.9)$$

where $\Phi^{k}_{(\lambda)}$ are the elastic distorsions (DJURIC [17]). When the free-energy function Ψ is introduced by $\Psi = \varepsilon - \theta\eta$ and

$$\Psi = \Psi(\Phi^{k}_{(\lambda)}, d^{k}_{(\lambda);K}, \theta, X), \qquad (7.10)$$

in accordance with the discussions of the section 1. the Clausius-Duhem inequality will reduce to

$$-\rho\dot{\Psi} + t_{i}^{j}\Phi_{j}^{(\lambda)}\dot{\Phi}^{l}_{(\lambda)} + h^{(\lambda)ij}\dot{d}_{(\lambda)i,j} - \rho\dot{\theta}\eta + \frac{1}{\theta}h^{i}\theta_{,i} \geq 0. \qquad (7.11)$$

The constitutive relations follow directly from this inequality

$$t_{i}^{j} = \rho\frac{\partial\Psi}{\partial\Phi^{i}_{(\lambda)}}\Phi^{j}_{(\lambda)},$$

$$h^{(\lambda)ij} = \rho\frac{\partial\Psi}{\partial d_{(\lambda)i;K}}x^{j}_{;K}, \qquad (7.12)$$

(7.12)
$$\eta = -\frac{\partial \Psi}{\partial \vartheta} .$$

In derivation of the relation (7.12) for the director stresses $h^{(\lambda)ij}$ it is assumed that the directors are affected only by the total deformation, which is compatible, and not by the the incompatible thermal (or elastic) deformation.

From the principle of objectivity for the function Ψ it follows that Ψ is an arbitrary function of the non-holonomic tensors

(7.13)
$$\overset{E}{C}_{\lambda\mu} = g_{lm}\Phi^{l}_{(\lambda)}\Phi^{m}_{(\mu)} ,$$
$$F_{(\alpha)\lambda\mu} = g_{lm}\Phi^{l}_{(\lambda)}d^{m}_{(\alpha);n}\,\Phi^{n}_{(\mu)} ,$$

but according to (2.18)

(7.14)
$$\overset{E}{C}_{\lambda\mu} = C_{LM}\,\Theta^{L}_{(\lambda)}\,\Theta^{M}_{(\mu)} ,$$
$$F_{(\alpha)LM} = F_{(\alpha)LM}\Theta^{L}_{(\lambda)}\,\Theta^{M}_{(\mu)} ,$$

where C_{LM} is the material tensor of the total deformation, and $F_{(\alpha)LM}$ are the director-deformation tensors

(7.15)
$$F_{(\alpha)LM} = g_{lm}\,x^{l}_{;L}\,F^{m}_{(\alpha);M} .$$

Further discussion of the elastic generalized Cosserat continuum is analogous to the preceeding discussion of the "ordinary" materials. The linearized constitutive equations in the special case, when the directors represent rigid triads ("micropolar" media), were extensively discussed in a series of papers by NOWACKI for detailed references see the list of publication in the Nowacki Anniversary Volume [15]).

In the theory of materials of grade two the internal energy function is considered in the form

$$\varepsilon = \varepsilon(x^k_{;K}, x^k_{;KL}, \eta, X) , \tag{7.16}$$

and when the stresses are produced by the (incompatible) thermal strains the deformation gradients have to be replaced by the distorsions and distorsion gradients. (STOJANOVIC [38])

$$\varepsilon = \varepsilon(\Phi_{(\lambda)}, \Phi^l_{(\lambda),k}, \eta, X) . \tag{7.17}$$

The energy equation in this case reads

$$\rho\dot{\varepsilon} = t^{(ij)} d_{ij} - m^{ijk}\omega_{ij,k} + \rho q + h^i_{,i} , \tag{7.18}$$

where d_{ij} and ω_{ij} are "incompatible" tensor of rate of strain and vorticity (cf. (4.10))

$$d_{ij} = \dot{\Phi}^k_{(\lambda)}\Phi^{(\lambda)}_{(i} g_{j)k} , \quad \omega_{ij} = \dot{\Phi}^k_{(\lambda)} g_{k[i}\Phi^{(\lambda)}_{j]} . \tag{7.19}$$

The Clausius - Duhem inequality becomes

$$(7.20)\qquad -\rho\dot{\Psi} + t^{(ij)} d_{ij} - m^{ijk}\omega_{ij,k} - \rho\eta\dot{\theta} + \frac{1}{\theta} h^i \theta_{,i} \geq 0 \ .$$

If we assume the free-energy function to be of the form

$$(7.21)\qquad \Psi = \Psi(\Phi^l_{(\lambda)}, \Phi^l_{(\lambda),k}, \theta, X)$$

the inequality (7.20) will be satisfied by

$$(7.22)\qquad t^{(ij)} = \rho g^{il}\left(\frac{\partial\Psi}{\partial\Phi^l_{(\lambda)}} \Phi^j_{(\lambda)} + \frac{\partial\Psi}{\partial\Phi^l_{(\lambda),k}} \Phi^j_{(\lambda),k} - \frac{\partial\Psi}{\partial\Phi^k_{(\lambda),j}} \Phi^k_{(\lambda),l}\right)$$

$$(7.23)\qquad m^{ijk} = -\rho g^{il} \frac{\partial\Psi}{\partial\Phi^l_{(\lambda),k}} \Phi^j_{(\lambda)} \ ,$$

$$(7.24)\qquad \eta = -\frac{\partial\Psi}{\partial\theta} \ .$$

The right- and left-hand sides of tensorial equations must possess the same tensorial symmetries. This requires that

$$(7.25)\qquad \left[g^{il}\left(\frac{\partial\Psi}{\partial\Phi^l_{(\lambda)}} \Phi^j_{(\lambda)} + \frac{\partial\Psi}{\partial\Phi^l_{(\lambda),k}} \Phi^j_{(\lambda),k} - \frac{\partial\Psi}{\partial\Phi^k_{(\lambda),j}} \Phi^k_{(\lambda),l}\right)\right]_{[ij]} = 0 \ ,$$

$$(7.26)\qquad \left(g^{il} \frac{\partial\Psi}{\partial\Phi^l_{(\lambda),k}} \Phi^j_{(\lambda)}\right)_{(ij)} = 0 \ .$$

The first of these conditions coincides with the system of e- quations which follow from the principle of objectivity when the function is required to be invariant under an arbitrary infinitesimal rigid rotation of the spatial system of reference. The conditions (7.25) and (7.26) are identically satisfied by

$$\overset{E}{C}_{\lambda\mu} = g_{lm}\,\Phi^{l}_{(\lambda)}\,\Phi^{m}_{(\mu)}\ ,$$

$$\overset{E}{D}_{\lambda\mu\nu} = [g_{lm}\,\Phi^{l}_{(\lambda)}\Phi^{m}_{(\mu),n}\,\Phi^{n}_{(\nu)}]_{[\lambda\mu]}\ , \tag{7.27}$$

and Ψ is an arbitrary function of $C_{\lambda\mu}$, $D_{\lambda\mu\nu}$, Θ and X. Using (2.18) the tensor $C_{\lambda\mu}$ and $D_{\lambda\mu\nu}$ may be related to the corresponding material tensors C_{LM} and D_{LMN},

$$\overset{E}{C}_{\lambda\mu} = C_{LM}\,\theta^{L}_{(\lambda)}\,\theta^{M}_{(\mu)}\ ,$$

$$\overset{E}{D}_{\lambda\mu\nu} = D_{LMN}\,\theta^{L}_{(\lambda)}\,\theta^{M}_{(\mu)}\,\theta^{N}_{(\nu)}\ , \tag{7.28}$$

$$D_{LMN} \equiv g_{lm}\Phi_{(\lambda)}\Phi^{m}_{(\mu)}\Phi^{n}_{(\nu)}\,\theta^{(\lambda)}_{[L}\,\theta^{(\mu)}_{M]}\,\theta^{(\nu)}_{N} = -\,D_{LMN}\ .$$

In the quoted paper Ψ was considered in the form

$$\Psi = \Psi(C_{LM},\, D_{LMN}\,,\, \Theta) \tag{7.29}$$

and the relations for stress and couple stress derived from Ψ in this form reduce in the linearized form for isotropic materials to the equations also considered by NOWACKI [29] .

In the theory of elastic materials of grade two

the couple-stress tensor m^{ijk} is not completely determined since in the energy equation (7.18) we have the term

$$(7.30)\qquad m^{ijk}\omega_{ij,k} = m^{ijk}\, v_{i,jk} = m^{i(jk)} v_{i,jk} \ ,$$

because of the symmetry of the velocity gradients $v_{i,jk} = v_{i,kj}$. Therefore only eight out of nine components $m^{i(jk)}$ of the tensor $\underset{\sim}{m}$ are satisfied by the inequalities (7.20). However, if the deformations do not satisfy the compatibility conditions, from (7.23) follows that all nine components of the couple-stress tensor are determined through the constitutive relations. Writing

$$(7.31)\qquad m^{ijk} = \rho \frac{\partial \varepsilon}{\partial D_{KLM}} x^i_{;K}\, x^j_{;L}\, x^k_{;M} \ ,$$

we obtain nine relations for the components of $\underset{\sim}{m}$. Assuming that the deformation is compatible, i.e.

$$(7.32)\qquad \theta^{(\lambda)}_L = \delta^\lambda_l \ , \quad \Phi^l_{(\lambda)} = x^l_{;\lambda} = \frac{\partial x^l}{\partial X^\lambda} \ ,$$

the tensor $\underset{\sim}{D}$ will reduce to

$$D_{KLM} = C_{M[K,L]} \ ,$$

which has only eight independent components (TOUPIN [45]), but all nine components of $\underset{\sim}{m}$ in (7.31) will remain determined (STOJANOVIC [39]).

References

[1] J.E. Adkins, A.E. Green, R.T. Shield, Phil. Trans. R. Soc., A 246, 181-213 (1953).

[2] D. Blagojevic, "A contribution to the nonlinear theory of instationary thermoelasticity" (In Serbo-Croat). Proc. XI Yugoslav Congress of Mechanics, Basko Polje, June 1972 (in print).

[3] L. Brillouin, "Les tenseurs en Mécanique et en Elasticité" Mosson, Paris, 1938

[4] L. Brillouin, "On thermal dependence of elasticity in solids" Phys. Rev. 54, 916-917 (1938).

[5] L. Brillouin, "On thermal dependence of elasticity in solids" Indian Acad. Sci. (A)8, 251-254 (1938).

[6] L. Brillouin, "On thermal dependence of elasticity in solids" Phys. Rev. 55, 1139 (1939).

[7] L. Brillouin, "Influence de la temperature sur l'élasticité d'un solide" Mem. Sci. Math. pass. 99 Gauthier-Villars, Paris, 1940.

[8] P. Chadwick, "Thermoelasticity. The dynamical theory" Progress in Solid Mechanics Vol. I (ed Sneddon and Hill), pp. 263-328. North Holland, Amsterdam, 1960.

[9] P. Chadwick, L.T.C. Seet, "Second-order thermoelasticity theory for isotropic and transversely isotropic materials" Trends in Elasticity and Thermoelasticity (W. Nowaki Anniversary Volume) pp. 29-57 W. Nordhoff,

Groningen, 1971.

[10] H.R. Chaudry, "A note on second-order effects in plane strain thermoelasticity", Int. J. Engng. Sci. 9, 673-678 (1971).

[11] B.D. Coleman, "Thermodynamics of materials with memory", Arch. Rat. Mech. Anal. 17 (1964).

[12] B.D. Coleman, W. Noll, "The thermodynamics of elastic materials with heat conduction and viscosity", Arch. Rat. Mech. Anal. 13, 167-178 (1963).

[13] B.D. Coleman, W. Noll, "Material symmetry and thermostatic inequalities in finite elastic diformations", Arch. Rat. Mech. Anal. 15, 87-111, 1964.

[14] B.D. Coleman, V.J. Mizel, "Existence of caloric equations of state in thermodynamics", J. Chem. Phys. 40, 1116-1125, (1964).

[15] 'Trends in Elasticity and Thermoelasticity" Witold Nowaki Anniversary Volume. (Editors R.E. Czarnota-Bojarski, M. Sokolowski, H. Zorski). Wolters-Noordhoff, Groningen, 1971.

[16] W.A. Day, M.E. Gurtin, "On the simmetries of the conductivity tensor and other restrictions in the nonlinear theory of heat conduction" Arch. Rat. Mech. Anal. 33, 26-32, (1969.

[17] O.W. Dillon, Jr., "A nonlinear thermoelasticity theory", J. Mech. Phys. Solids 10, 123-162, (1969).

[18] S. Djuric, "Thermoelasticity of generalized Cosserat continuum", (in Serbo-Croatian). Proc. X Yugoslav Congress of Mechanics, Basko Polje, 1970, pp. 135-151. Belgrade 1972.

[19] J.L. Ericksen, "Tensor fields", Hadnbuch der Physik

Bd. III/1 (ed. S. Fluegge), pp. 794-585, Berlin, 1960

[20] A.E. Green, J.E. Adkins, "Large elastic deformations and nonlinear continuum mechanics", Oxford Univ. Press. 1960.

[21] A.E. Green, "Thermoelastic stresses in initially stressed bodies", Proc. Roy. Soc. London, Ser. A 266, 1-19, (1962).

[22] A.E. Green, N. Laws, "On the formulation of constitutive equations in thermodynamical theorie of continua", Quart. J. Mech. Appl. Mech. 20, 265-275, (1966).

[23] G. Grioli, "Mathematical theory of elastic equilibrium, (recent results)", (Eng. angew. Math. 7), Springer, Berlin-Gottingen-Heidelberg, 1962.

[24] R.A. Groot, "Thermodynamics of a continuum with microstructure', Int. J. Engng. Sci. 7, 801-814, (1969.

[25] G. Herrmann, "On second-order thermoelastic effects', ZAMP 15, 253-261, (1964).

[26] F. Jindra, "Warmespannungen nichtlinearen Elastizitatsgesetz", Ing. Arch. (1959).

[27] A.F. Johnson, "Some second-order effects of thermoelasticity theory", Report DNAM 84, National Physical Laboratory, 1970

[28] W. Nowacki, "New trends of investigation in thermoelasticity", Plater and shells (ed. J. Brill and J. Balas), pp. 71-88, Slovak Acad. Sci., Bratislava, 1966.

[29] W. Nowacki, "Couple-stresses in the theory of thermoelasticity", I,II,III. Bull. Acad. Pol.

Sci. 14, 97-106, 203-212, 505-513, (1966).

[30] H. Parkus,"Thermoelasticity", Blaisdail, Waltham (Mass). 1968.

[31] A.C. Pipkin, R.S. Rivlin, "The formulation of constitutive equations in continuum physics", Div. Appl. Math. Brown Univ. Report Sept. 1968.

[32] J.A. Schouten, "Ricci-Calculus", (2nd edition), Springer-Verlag, Berlin-Gottingen-Heidelberg, 1954.

[33] L.I. Sedov, "Introduction to the mechanics of continuous media", (in Russian), Gosud. Izdat, Fiz-Mat, Lit. Moskow, 1962.

[34] A. Signorini,"Sulle deformazioni termoelastiche finite", Proc. 3rd Int. Congr. Appl. Mech., Stockholm 2, 80-89, (1930).

[35] A. Signorini, "Trasformazioni termoelastiche finite", Annali di matematica, 22, 33 (1943), 30, 1-72 (1949), 39, 148-199 (1955), 51, 329-372 (1960).

[36] R. Stojanovic, "On the stress relation in non-linear thermoelasticity", Int, J. Non-Li,ear Mech. 4, 217-233 (1969).

[37] R. Stojanovic, "Mechanism of polar Continua", Courses and lectures N° 2. CISM, Udine 1969.

[38] R. Stojanovic, "A non-linear theory of thermoelasticity with couple-stresses", Trends in Elasticity and Thermoelasticity, (W. Nowacki Anniversary Volume), pp. 249-266, W. Noordhoof, Groningen, 1971.

[39] R. Stojanovic, "A contribution to the theory of polar elastic materials", Proc. X Yugoslav Congress

on Mechanics, Basko Polje 1970, pp. 559-570, Belgrade, 1972.

[40] R. Stojanovic, "Recent developments in the theory of polar continua", Courses and lectures N° 27, CISM, Udine, 1970, Springer Verlag Wien-New-York.

[41] R. Stojanovic, S. Djuric, L. Vujosevic, "On finite thermal deformations", Arch. Mech. Stosow. 16, 103-108 (1964).

[42] R. Stojanovic, L. Vujosevic, S. Djuric, "On the stress-strain relations for incompatible deformations', Theory of Plates and Shells (ed. by J. Brilla and J. Balas), pp. 459-467, Slovak Akad. Sci. Bratislava, 1966.

[43] C. Tolotti, "Sul potenziale termodinamico dei solidi elastici omogenei e isotropi per trasformazioni finite", Mem. dell'Accad. d'Italia 14, 529-541 (1943).

[44] R.A. Toupin, "The elastic dielectric", J. Rat. Mech. Anal. 5, 849-915, (1955).

[45] R.A. Toupin, "Elastic materials with couple-stresses", Arch. Rat. Anal. 11, 385-414, (1962).

[46] R.A. Toupin, "Theories of elasticity with couple-stress", Arch. Rat. Mech. Anal. 17, 85-112, (1964).

[47] C. Truesdell, "The mechanical Foundations of elasticity and fluid dynamics", J. Rat. Anal. 1, 125-171, 173-300 (1952).

[48] C. Truesdell, "Rational thermodynamics", McGraw-Hill, New-York, 1969.

[49] C. Truesdell, W. Noll, "The nonlinear field theories of mechanics", Handbuch der Physik, Bd.III/3

(editor S. Fluegge). Berlin-Heidelberg-New-York, 1965.

[50] C.C. Wang, "On the symmetry of the heat conduction tensor', Appendix 1 to C. Truesdell: Rational thermodynamics, McGraw-Hill, New-York, 1969.

[50] C.C. Wang, R.M. Bowen, "On the thermodynamics of non-linear materials with quasi-elastic response", Arch. Rat. Mech. Anal. 22, 79-99, (1966).

[52] C. Wozniak, "Thermoelasticity of non-simple oriented materials", Int. J. Engng. Sci. 5, 605-612, (1967).

[53] C. Wozniak, "Thermoelasticity of bodies with micro-structure", Arch. Mech. Stosow. 19, 335-365, (1967).

[54] C. Wozniak, "Theory of thermoelasticity of non-simple materials", Arch. Mech. Stosow. 19, 485-493, (1967).

Contents

Printed in the United States
By Bookmasters